THE BIRTH OF A NEW AGRICULTURE

Koberwitz 1924

Compiled and edited by Adalbert Graf von Keyserlingk

TEMPLE LODGE
London

Translated by John M. Wood

Temple Lodge Publishing
51 Queen Caroline Street
London W6 9QL

First English edition published by Temple Lodge 1999

Originally published in German under the title *Koberwitz 1924, Geburtsstunde einer neuen Landwirtschaft* by Verlag Hilfswerk Elisabeth, Stuttgart, in 1974

© Verlag Hilfswerk Elisabeth, Stuttgart 1974
This translation © Temple Lodge Publishing 1999

The moral right of the translator has been asserted under the Copyright, Designs and Patents Act, 1988

All rights reserved. No part of this publication may be reproduced, stored in a retrieval system, or transmitted, in any form or by any means, electronic, mechanical, photocopying, recording or otherwise, without the prior permission of the publishers

A catalogue record for this book is available from the British Library

ISBN 1 902636 07 4

Cover by S. Gulbekian. Cover photo of Koberwitz courtesy of David Clement
Typeset by DP Photosetting, Aylesbury, Bucks.
Printed and bound in Great Britain by 4edge Limited, Essex

*The publishers dedicate this edition to
John M. Wood (1913–1998)
who died in Fort William, Scotland, shortly after finishing the translation of this book. (He had been attending a conference on the Isle of Mull.)*

Contents

Publisher's Note	vii
Foreword by Rudolf Meyer	1
Foreword to the 1974 edition	8

PART I: Twelve Days with Rudolf Steiner
Revised new edition from the posthumous writings of
Countess Johanna Keyserlingk

Foreword to the edition of Easter 1949	15
Foreword to the English edition of 1952	16
My first encounters with Rudolf Steiner	20
Memories of Eliza von Moltke	30
Encounters 1919–1921	35
Count Carl von Keyserlingk	46
Rudolf Steiner's first visit to Koberwitz 1922	52
Rudolf Steiner's second visit to Koberwitz 1922	58
The Agricultural Course 1924	60
Esoteric conversations	81
Concluding notes by Count Adalbert Keyserlingk	95

PART II: Accounts by those who were present

Count Alexander von Keyserlingk	99
Paula Eckardt	109
Luise von Zastrow	113
Günther Sponholz	117
Lutz Engel	119
Wilhelm Rath	129
Karin Ruths-Hoffman	134
Helmuth Woitinas	139
Erna van Deventer	144
Count Adalbert von Keyserlingk	150

PART III: The Whitsuntide Gathering in its
Historical setting. Concluding remarks by Count
Adalbert von Keyserlingk

    The historical stage: events between 1912 and 1924   163
    Karmic connections of the 4th and 9th Centuries   182
    The Koberwitz impulse as hope for the future   189

Appendix 1: Requiem by Rudolf Meyer (shortened)   219
Appendix 2: Letter from Rudolf Steiner to Count
    Keyserlingk, 1919   222
Illustrations   224
Notes on the sources and literature of Part III   227
Bibliography   230

Publisher's Note

Although this book is being given a full publication in English, it should be noted that its contents are intended primarily for those who feel connected to anthroposophy, as founded by Rudolf Steiner (1861–1925). While it contains a wealth of valuable material, the book also includes some passages which are not entirely clear, even in the original German. Nevertheless, we have taken the decision to leave the text largely unaltered as to structure and content, so that readers can evaluate the material for themselves. There is no doubt that there are many treasures in these unique accounts of Rudolf Steiner's efforts to establish a new, transformed agriculture for the modern age.

The Koberwitz conference was hosted by Countess Johanna von Keyserlingk and her husband Count Carl von Keyserlingk. The Countess's reminiscences make up Part I of the book. However, the book was edited, compiled and published in the original German by their son, Adalbert Graf von Keyserlingk (whose wife Gerda co-wrote the Foreword to the 1974 edition), and he also included his commentaries and essays which comprise Part III. The Count Alexander von Keyserlingk, whose account of the Koberwitz conference opens Part II, is the nephew of Count Carl von Keyserlingk.

Helmuth von Moltke, who is mentioned a good deal in this book, was Chief of the General Staff of the German army during the outbreak of the First World War. His wife, Eliza, was one of Rudolf Steiner's staunchest esoteric pupils. The important correspondence between Steiner and Eliza and Helmuth von Moltke (including the after-death communications from the latter)—referred to particularly in Part III—have now been published in English in the volume *Light for the New Millennium*. For publication information on this and all other books which are referred to in the text, see the Bibliography on page 230.

Know what lies at the heart
Of the world's spiritual basis
And it will teach you how to probe
Into your own soul powers.
Search outside of you for what is within
And search within for what is outside.

Des Innern Wesen erkenne
In den Welten-Geistes-Gründen,
Und der Welten Innenkraft,
Es kann sie dir verkünden
Das Forschen in eigner Seelenmacht.
So such im Äußeren das Innere
Und in dem Eigenen die Welt.

(Written by Rudolf Steiner in the visitors' book of Count and Countess Keyserlingk in Koberwitz. February 1922.)

Foreword

Koberwitz, an estate which lies near Breslau, gained special significance during the last years of Rudolf Steiner's life. Many friends inclined to anthroposophy received hospitality there, often for lengthy stays, and were able to feel perfectly at home in the unique atmosphere of this spacious, modern mansion. They became acquainted with a place which had a forward-looking spirit, balanced by a down-to-earth rural element which gave it a healthy human touch.

In my epilogue to *Gralburg* ('The Grail Castle'), the first volume of Countess Keyserlingk's posthumous writings, I described my first meeting with her. It was on the occasion of a teachers' conference in Breslau at Easter 1921. Dr Erich Schwebsch and I were invited to stay at Koberwitz for about a week as participants. My meeting with the Count was of no less importance to me at that time than my meeting with the Countess. One day we toured through a series of Silesian villages all belonging, as large estates, to the family concern of which Count Carl Keyserlingk was then in charge. Whatever subject he touched upon, whether dairy farming, sugar beet growing or human relationships on the estates, it was all accompanied by a down-to-earth familiarity with the subject discussed, and at the same time a consciousness of the urgent needs of our time in relation to Rudolf Steiner's social ideas. As a pupil of the spirit he was filled with deepest reverence and devotion for this, but that did not interfere in any way with the pre-eminence of his bearing which expressed itself in a simple kindheartedness and modesty.

When we founded the Christian Community in Dornach in Autumn 1922 and Rudolf Steiner advised me to go to Breslau, I was immediately assured of the sympathy and helpfulness of Count and Countess Keyserlingk. I was allowed to be their guest for three years and to carry out my work in Silesia from there. I do not remember ever having asked in vain for either inner or outer help for our young movement or for our friends in time of need. This companionship in the daily routine and the tasks which pressed upon us, provided the background out of which we could conduct our 'Grail conversations', as described in *Gralburg.*

During the first period of our activity, our worthy friend Rudolf von Koschützki and his wife were guests at Koberwitz Manor. Rudolf von Koschützki had been a farmer in Upper Silesia in his youth, then a writer of books of inspiring human interest, but also many books on agriculture. In his later years he found a relationship to our movement for religious renewal and was one of its founders. With Kurt von Wistinghausen and myself, he had taken over the office of priest for Silesia in the autumn of 1922. The three of us, together with the household at Koberwitz, celebrated the Act of Consecration of Man there, before the Community in Breslau was established at Advent.

Because of Herr von Koschützki's experience in agricultural matters, an immediate bond was formed with the Count which enabled him to share in the initiative which led to the Agricultural Course given by Rudolf Steiner in Koberwitz at Whitsuntide 1924. This formed the starting point for everything which developed later as the biodynamic movement. With this spiritual deed we could say that the name of Koberwitz is written into the annals of cultural history insofar as a new beginning was made in the face of western civilisation's decline. This is one of the fruitful cultural initiatives produced by Rudolf Steiner during the last years of his activity on earth. 'Europe is sitting on the edge of a volcano and does not realise it' he could say at that time out of his deep concern for the future of our western culture. Those who were able to be present at the conversations at meal times during the days of the conference (7–17 June 1924) often heard Rudolf Steiner refer to the storm brewing just then over the whole world as something unavoidable. He spoke about the destruction of Central European towns. 'The enemies of Germany will not be content until the last chimneys have been razed to the ground'—he was referring to the factory chimneys and added that a kind of universal justice was expressed therein, because nowhere in the world had modern industry produced such poisonous outgrowths as in Central Europe. None of these statements, however, had a paralysing effect; they aimed to kindle the spiritual courage of mankind. He simultaneously inspired people with hope when he spoke about the necessity of forming quiet centres of a new spiritual life in agrarian seclusion which could have far-reaching effects if the western nations would send their sons and daughters to them.

He also spoke at that time about the new 'wisdom of youth' which he wished to impart. He was conscious of the burden of responsibility which the coming generation would have to bear and he wanted to equip them with spirit-powers which would enable them to stand firm in the coming universal turmoil.

He spoke in this way on many occasions, at specially appointed times, to circles of young people with urgent questions. He once characterised the souls of those who had incarnated at the turn of the century as those who had the will to ensoul all earthly relationships. He mentioned, moreover, an individual who had the same urge in the 19th century: Friedrich Nietzsche, who belonged, like an elder brother, to this karmic community of souls which had now descended to earth. It was his tragic fate to enter earth-life prematurely, for at that time the new spirit-light had not yet dawned.

Without being distracted by the great upheavals which have inevitably broken over the 20th century, Rudolf Steiner continually planted the seeds of a future civilisation. For that is how we looked upon all that he inaugurated at that Whitsuntide festival in the way of new agricultural methods for healing the ailing earth, as part of an all-embracing therapy for humanity. How warmly and knowledgeably he spoke, for instance, about the way to build a compost heap! One was immediately transported to a farm setting and was forced to believe that this was the real job in which he felt at home. It seemed to be quite natural when Count Keyserlingk afterwards addressed Rudolf Steiner in the name of the landowners, farm workers and gardeners as the 'great peasant' from whom everyone present would gladly take instruction for their profession. Naturally Rudolf Steiner disclaimed this honour and insisted that, at the very most, he was just 'a small peasant' by virtue of the help he gave on a farm during his boyhood.

On such occasions many other things took place besides the daily lecture on agriculture and the hour for question time, which were the main purpose of the conference. A veritable cornucopia of spiritual gifts was showered upon us. We were offered a eurythmy performance in Breslau and a dramatic rendering of Goethe's *Iphigenie* by a group of actors who were hoping at that time that a course on speech formation and drama would be given by Rudolf Steiner. However, every evening we could still listen to a lecture about the laws of karma as the 'forming of destiny in human life', in which he described the after-death path of the human soul through the planetary spheres and its further progress from one life to the next in conjunction with the sublime spirits of the heavenly hierarchies. Goethe and Schiller, Voltaire, Heinrich Heine, Garibaldi, Harun al Raschid, Comenius and Pestalozzi were some of the historical personages illustrative of this world tour through the centuries. In those Whitsuntide days a humanity was expressed which knew how to unite loving devotion to earthly tasks with courage toward the stars.

Those working at that time in the movement for religious renewal who were allowed to take part in the courses, received another, special gift. Since the founding of the Christian Community in the autumn of 1922 all the festivals of the Christian year had been celebrated with the help of Rudolf Steiner; only the St John's festival had been left out. In our youth we had celebrated it as a pagan midsummer festival with its symbolic flame. Now it was to be created afresh in Christian fashion. Christ Himself, the being who descended to us out of the Sun- heights of the universe, wants to be received ever and again by souls in need of light, as He was once received into the sheaths of a man at the turning-point of time through the Baptism in Jordan. But John, the herald of this turning-point of time, was to be addressed in a significant manner during this celebration. The Genius of Christianity Himself seemed to want to take a hand once more in the leadership of the age. It was during a break in the agricultural lectures in Koberwitz that Rudolf Steiner approached us and quite simply, but with inner joy, handed us a paper on which was the ritual of the St John's Service, which he had completed that night.

During one of the youth gatherings which were so helpful to many of the young people, a confirmed believer in the 'youth movement' approached with a complaint about his comrades with whom he had once met Sunday after Sunday to ramble in the countryside, enjoy camp fires and sleep in the straw. 'But ever since my companions read *your* books' he said to Rudolf Steiner, half reproachfully and half in search of advice, 'they do not want to go rambling any more every Sunday, and our lovely "Wandervogel" community is falling apart'. He could not understand, said Rudolf Steiner, how youthful friendship and commonly held ideals should fade as a result of preoccupation with anthroposophical ideas. *'How To Know Higher Worlds* is a genuine ramblers' book; every sentence has been "walked through",' said its author. 'It is true however,' he went on to say, 'that after one has practised the exercises given in this book, one might reach the stage of stopping at the first wayside flower one comes across, for one can find the whole of the starry firmament hidden therein. Perhaps it is not necessary to "wander" so far if one discovers how to penetrate the heights and depths of the cosmos by means of simple phenomena.' And then, once more, he assured us: 'It has really been *walked through*; as such, it is not actually a book. One has to reach out beyond the printed page.'

The youth movement during the first decades of the 20th Century was full of fervour in its search for nature. It wanted to get away from city life in which the breath of approaching doom was beginning to be felt. It was with sadness that

Rudolf Steiner beheld the illusions which lay hidden therein, which lured dreaming souls towards the abyss. Thus, on the morning of departure, Rudolf Steiner gave an address to a small group of young people at Koberwitz mansion, after having done a new deed in the service of the earth through his Agricultural Course, and thereby pointed the way to a true priesthood of nature.

His words of warning were impressive when he spoke about Rousseau, whose 'Back to Nature' doctrine had so intoxicated European youth of the 18th century. As this call went forth, he said, one could hear the snickering of the demons behind the scenes of outer happenings. This snickering rose to derisive laughter, however, after this nature-worship turned into the materialism of natural science in the 19th century. The soul is in danger of losing its humanity today, he said, if it blindly follows nature before having discovered the living spirit within itself. One has to grasp in living pictures what is the innermost striving of youth. Two words have joined together to become a symbol for the strivings of youth today. They are the words 'Wander-Vogel', [literally: 'Wandering Bird', the name for the German youth movement].

Does humanity today have any idea what *Wandern* meant in olden days? What going on a 'pilgrimage' meant? We must go back to the times of the primeval wanderer who once met souls in wind and storm and revealed to them the secrets of the world: that was Wotan, the World-Wanderer. Only when one had met this wanderer face to face did one really know what it meant to 'wander'. And who can know nowadays, when confronted by the world of birds, that one has to go through the experiences that Siegfried had which enabled him to understand the language of the birds? Only one who can wield the Siegfried-Sword which is able to slay the dragon, the sword which is only a prophetic forerunner of Michael's sword, can also become worthy of learning the hidden speech of the birds. He who is prepared to follow this twofold path which leads to Wotan and to Siegfried, earns the right to use the double name 'Wander-Vogel'.

Through such instructions given by Rudolf Steiner one received intimate impressions about which one hardly dares to speak. It was striking to sense that he was completely transformed into that of which he was speaking. As though he were immeasurably old, coming out of long past ages, with weathered features and breathed upon by winds of long ago, he appeared to us at that moment as the primeval wanderer himself. Wotan was there in person before us. Shortly afterwards he changed into something completely different as he spoke about Siegfried, the Dragon-Slayer: he was a youth, at the sight of whom one could conjecture how young the spirit, the Holy Spirit, could make one. At that moment he became a witness to the dawn of humanity's youth.

After this address with which he ended his last Whitsuntide festival on earth, he left the Silesian estate. He had to get to Jena that same night, wishing to visit the site of a special deed of love to which a number of young people felt drawn. This was the Lauenstein, where the curative movement started. Thus, immediately after giving the Agricultural Course, Rudolf Steiner founded what, in decades to come, was to bring the greatest blessing to humanity, the curative education work.

It was a storm of activity which called Rudolf Steiner at that time from one piece of work to another; he created new professions and showed new goals to young people willing to accept the spirit. The Koberwitz impulse is one of those germinal deeds which will come to their full fruition only after the end of this millennium.

Rudolf Meyer

Foreword to the 1974 German Edition

We have continually received requests to re-publish the volume *Twelve Days with Rudolf Steiner*. But there were also many dissenting voices: some thought these reminiscences were too personal, and belittled the great initiate by describing everyday happenings. Others, though, especially young people, particularly wanted to hear these more everyday things about Rudolf Steiner and were pleased to know that he entered fully into his surroundings and into the demands and social life of his time.

A further criticism was that by describing Rudolf Steiner in this way we were only describing ourselves and that it is very presumptuous to disregard all that other people had experienced. Such comments as these were made chiefly by those who could not make up their minds to take up the pen themselves. There is a difference, however, between giving an objective picture of someone's life with exact dates and places, and describing personal experiences. The latter can only occur in the mirror of one's own soul: 'How did I see the other person? What were my impressions when I encountered him?'

Painters or poets can produce a coloured or sounding image of friend and foe, but all of us—whether peasant or man of letters—experience the stirrings in our own soul which respond through colour and sound when we meet another person who is able to rouse the colour and sound within us. Thus we welcomed every contribution which bore the features of the writer's own experience, which embodied those things which the meeting with Rudolf Steiner had impressed at that time, 1924, on each person's mind as an imprint of true humanity.

We sought out those people whom Rudolf Steiner met when the Agricultural Course was being given. Regrettably—after nearly 50 years—there were only very few of them. If it had been ten years earlier a more colourful picture would have been produced! But we have the descriptions in 'Rudolf Steiner by his pupils' (*Golden Blade* 1958) and other books belonging to this category, the Agricultural Course itself, the addresses given to young people, and the News-Sheets.

What is described here is not meant to be a contribution to a history of the bio-dynamic movement—if that were so, many points of view would be lacking. It is a record of personal recollections, albeit one which shows the full range of subjects which were dealt with in Koberwitz.

Another serious consideration in respect of this work is the fact that it contains esoteric material of a personal nature. Nevertheless, Rudolf Steiner often reiterated the fact that, since the Christmas Foundation Meeting, concealment of certain esoteric matters was no longer justified. The trustees of Rudolf Steiner's Estate (Nachlaßverwaltung) have also since then published meditative material which was originally intended only for certain karmic groups. It is therefore not just a matter for the authors to decide, but also for the readers to discriminate as to what use they make of it. All those who have contributed to this work, but especially the author of *The Twelve Days*—and some had even received this as a task from Rudolf Steiner—wished to impart the experiences, events and information which they had received on their path of spirit pupilship, to those who could receive them with due respect. For nowadays the one who receives is also responsible for the use he makes of what he hears and reads. And many of the readers of *The Twelve Days* have reported soul experiences they have had as a result of what was contained in it. On the basis of the above-mentioned objections, however, we have decided to preface this book with a notice to the effect that this volume is for those with a basic knowledge of anthroposophy, who are already familiar with anthroposophical terminology, for only so can they properly understand the esoteric information here given.

We would like to add that the author of *The Twelve Days* had 28 instruction sessions with Rudolf Steiner, which are being published as single leaflets under the title: 'From Countess Johanna von Keyserlingk's posthumous writings' the first two of which: *Gralburg* ('Grail Castle') volumes 1 and 2, and the third volume: *Erlöste Elemente* ('Redeemed Elements') have already appeared in print. In the present work the articles and questions relating to the Koberwitz impulse have been included. In the course of this acccount it will become evident that it was only with the help of Rudolf Steiner that the Koberwitz estate came into the possession of the Keyserlingk family, who took up residence there four years before the course was given and left it four years after it ended. Thus the Agricultural Conference with its mighty impulses was able to take place in this secluded and personal setting. Immediately after the course had taken place the forces of opposition got to work. Through the opponents of this new agriculture—Professor Burgk and members of the Supervisory Board of the family company (connected with the artificial fertiliser industry)—the real purpose of the Koberwitz estate was taken away from it again, after this circle of people had been thwarted in their attempt to buy a copy of the newly printed Agricultural Course from Count Keyserlingk for a million [marks]. The first copies of this book were numbered and were only issued to anthroposophical farmers in exchange for a signed guarantee that the person receiving it would keep the contents secret until exact results of the new agricultural methods had been established.

Because everything is so quickly forgotten nowadays, it seemed to us to be necessary to describe what happened between the years 1912 and 1924. That influenced the way in which Rudolf Steiner was able to bring about the birth of a new agriculture.

Above all else the personalities of the two Moltkes were important for Rudolf Steiner because, through them, he was able to bring certain impulses to mankind. He always needed human beings and their destinies to be able to unite new things with old traditions. Thus the site of Koberwitz was decisive and the destinies of the 'Iron Count' and the 'Iron Countess' were necessary. Eliza and Helmuth von Moltke, with their particular karma, enabled the door for certain impulses from the spiritual world to be opened.

When we consider this fact we find that opponents of these personalities were necessarily also involved. It was for this reason that Part III of this book was written. It is intended to show, to begin with, how the social structure in 1917/18 was changed by certain people: it is, after all, people who make history. The observations contained there have been confirmed by my research into the Michael Sanctuary in Monte Cargano and, in conjunction with that, my study of the 4th and 9th centuries. This research was able, at least to some small extent, to justify Rudolf Steiner's suggestion that the causes of what happened in 1914–1918 should be sought in those earlier centuries.

Thirdly, we felt we had an obligation to recognise and pay tribute to the almost superhuman efforts of Rudolf Steiner amidst the chaos of the aftermath of the First World War. Part III is intended to show, therefore, how the Koberwitz impulse was the last Whitsuntide message of the great initiate. After many personal sacrifices by Rudolf Steiner, this agricultural impulse came about and showed itself to be the potential salvation of Central European culture.

In this new edition it has been found necessary to alter the style here and there, because some entries were in a diary-form dating from the beginning of the century. This diary-form was abandoned and the sequence of events was adapted accordingly. And because many contributions concerning the Koberwitz Whitsuntide gathering had in the meantime been made public, we were able to omit some of the personal recollections and earlier recordings. These were only included insofar as they seemed necessary to make things understandable. One might criticise as superfluous the word for word citing of talks in which Rudolf Steiner referred to 'iron' and the 'Iron Count' five times. But to us it seems that this was not merely done out of politeness or in jest. Rudolf Steiner knew quite well what he was doing, for iron, connected with the mysteries of the blood's Michaelic power of consciousness, is one of the central themes of the Koberwitz impulse. That is demonstrated also by arguments within the bio-dynamic movement about the use of artificial fertiliser on iron-rich Podsol earth. About iron and its Michaelic mission Rudolf Steiner said on another occasion: 'Thou mouldest it to thy service—yet it will only bring healing when to thee is revealed the lofty power of the spirit.'

This Whitsuntide message of Rudolf Steiner's was dedicated to the healing of the earth, and the Koberwitz impulse becomes a Michaelic task for those who feel united with iron in this way.

At this point we would like to thank all those friends who have taken the trouble to write accounts. Some of these were conversations first recorded on tape, while others arrived ready for the press. We should also like to thank those who gave us advice after having read the first edition of *Twelve Days with Rudolf Steiner*; and quite special thanks are due to those who helped with careful editing of the manuscript, to bring it in line with present day literary style, and who assisted with the publication.

Countess Gerda Keyserlingk and
Count Adalbert Keyserlingk

PART I

Twelve Days with Rudolf Steiner

Revised new edition from the posthumous writings
(IV, V) of Countess Johanna von Keyserlingk

Foreword to the Easter 1949 Edition

A quarter of a century has gone by since the memorable days of the Koberwitz Course; and now, in response to requests, I have decided to make my notes available to others.

These notes were intended only for my family, and on this account the very personal element which they contain must be excused. And yet how else should a diary be written than from one's own personal experience?

I met with so much delight in those who read its pages because they felt to be included in this daily life around Rudolf Steiner. And when I considered how few people still remain who can pass on to the younger generation a memory of this figure—who was pre-eminent even in the small details of daily life—it seemed to me right to make these pages accessible to wider circles. One youthful reader of the Diary said to me: 'We can read the books and lectures of Rudolf Steiner, but we cannot experience Rudolf Steiner the man. Some things could perhaps have been omitted yet the little actions of everyday life, from which one can read a person's nature, are what we of a later generation can no longer experience. These are the things that interest us.'

Another inducement to publication was the feeling of responsibility which I involuntarily bear, since important impulses given to younger generations—in whose hands the future rests—were spoken in our house, and therefore to some extent placed under our protection.

The reports of the Youth Conference, as also of the lectures of 7–17 June, 1924, are in order, inasmuch as they were prepared in my house, that is in my presence. They are here published in their original form.

Stuttgart, Easter 1949
Countess Johanna von Keyserlingk

Foreword to the English Edition of 1952

The publication of my Diary was a first experiment—I wanted to see how it would be received, and I am grateful to the friendly reception that has been given to it. Yet there remains still more of what Rudolf Steiner gave us there, in our lost Eastern Germany, and I feel a deep responsibility not to keep this to myself.

The concept of a threefold task, of worldwide significance, arising out of the Koberwitz Course, lived in the thoughts of Carl Keyserlingk:

1. Spiritualising the sciences,
2. Sanctifying the earth,
3. Helping to counteract the worldwide danger threatening man's nutrition

Very earnest words concerning the world's economy were spoken by Rudolf Steiner on 20 June 1924, three days after his return to Dornach—at a time, therefore, when he was still imbued with the experience of his days in Koberwitz. He said that not only was the moral development of humanity degenerating in the present transition to the Light Age, but that what man has made out of the earth itself and out of what lies immediately above it is also in a state of rapid degeneration. Rudolf Steiner went on to say that the forces which must be drawn forth from the spirit are still quite unknown today. Through the discovery of these forces the life of man in the physical sense could then continue on earth.

It is to be seen precisely in the sphere of agriculture, he said, that what proceeds from the spirit can be put to practical use. The whole heavens, for instance, with the stars, are involved in plant growth. People must know this, it must become part of their knowledge.

No one today is really aware of what manure signifies for the soil or how it should be used. Nor is it known how the mineral manures contribute to the deterioration of food products. The actual secrets of manure belong to the most interesting, most extraordinary mysteries.

Perhaps, through a high degree of spirituality, it might be possible for life itself to be directly influenced, so that through the right practical methods spiritual forces could enter and take hold.

Very few people today know that all the products upon which man lives have been deteriorating during the last three decades. This degeneration is a fact that can be statistically analysed, and on the basis of which it can be calculated that in a few decades produce will no longer be able to serve the nourishment of mankind. Through this degeneration our whole nutrition is called into question.

These are indications given by Rudolf Steiner in Dornach in June 1924. His statements are borne out year by year and have become still more serious in view of the immense increase in the earth's population.

It was this sphere of world-economy which Carl Keyserlingk had in mind and which was already playing its part.

The second task which lay close to his heart is the 'sanctifying of the earth'.

Spiritual science teaches us that when the blood flowed to earth from the Cross on Golgotha, a new Sun-globe was born in the interior of the earth. And it is there that the Christ, the Genius of the Sun, has since His descent to earth made His throne, as Regent of the earth.

If we take this knowledge seriously it can become a fact to us that golden rays can ascend from this centre, that they can break through the earth-crust and stream into the plants. These are the new currents of life by which plants may be regenerated and even appear in a new form.

It would help those who till the ground if they knew that the earth which bears the plants can become holy if man carries in his consciousness this knowledge of the throne of Christ in the golden centre of the earth's interior.

I do not know if Rudolf Steiner has referred to this so concretely anywhere else. His words at Koberwitz, telling us that this golden earth-centre is the legendary land of Shambhalla, sunk away from humanity and to be rediscovered by the seeking, Christ-guided human soul, are of great importance.

This should be borne in mind not only by those who cultivate the earth, but also by the human soul seeking for the heart of the divine. It is a great help to know that the light of the spirit radiates down to us from the heights, and yet that earthly man is also carried by the eternal divine world of the depths. The soul that seeks God may know that an actual union exists with the divine source from which the soul has sprung.

Rudolf Steiner spoke wonderful words to me of the divine heart of the Nazarene in the depths of the earth, which hears all that stirs the human soul, where all human sorrow and human joy are received and the prayer of the petitioning soul is accepted.

> May there ascend from the Depths
> The prayer that is heard in the Heights

are words from the Christmas Meditation of 1923. We may therefore pray to the heart in the earth's depths. Thence come helping rays for earth and mankind.

There is still a third thing of which I should like to speak in this Foreword, even though it may seem that one should not speak of it in the same breath as what has gone before.

Rudolf Steiner spoke of the iron in the Koberwitz soil and its will-energies. He thereby hinted at the still uninvestigated mystery of iron and the 'lofty power of its spirit'.

The earth underwent a change through the deed of redemption on Golgotha, a transformation of the unruly Mars powers of iron into the silvery light of Mercury.

Seen from this standpoint a new meaning is given to Rudolf Steiner's oft repeated reference to the 'Iron Count'. This was no jest but points to the fact that the devout farmer tilling his soil with a feeling of responsibility may play his part in this transformation. This allusion has not been understood and thus has generally been omitted from the printed publications. It was really a reference to the future mission of iron. To consciously carry within himself this 'sanctifying of the earth', to co-operate in this transformation, was the aspiration of the 'Iron Count' , one which already filled him as he stood before Rudolf Steiner in Berlin in 1918. This transforming of iron through the sacred mystery of the Christ-blood is a determining factor in the transformation of earthly substance, which will later be carried over into the 'transubstantiation of the earth'.

That was the underlying motif of the 'Iron Count' who looked on the Agricultural Course as forming a stepping-stone to this ultimate aim. His soul was so filled with this thought, and he referred to it so repeatedly in the last days of his life, that one could do no other than inscribe these words of his upon his urn:

> To create centres of peace and love
> in which the Christ can resurrect

Countess Johanna Keyserlingk

My first encounters with Rudolf Steiner

Berlin, 24 July 1918

I had heard Rudolf Steiner speak at two evening lectures and had been introduced to him on those occasions. But now I was to experience something quite special: her Excellency Eliza von Moltke had invited me to her house; there I was to meet Rudolf Steiner to tell him, at her request, about my life up until then.

Full of expectation I proceeded along the Schillerstraße in Charlottenburg in order to arrive at the house punctually at 4 pm. A minute after my arrival the doorbell rang and Rudolf Steiner entered. I was amazed to find that such a busy and famous man was so punctual. He was shown into the study, whereas I waited in the sitting room.

Soon Frau von Moltke entered with Rudolf Steiner and we took our places at the neatly set tea table. As well as tea, cherries and Swiss cheese had been laid for Dr Steiner.

Now, for the first time, I was able to experience the wonder which gripped one in the fascinating and dignified presence of Rudolf Steiner. At first only topical subjects were discussed, for Frau von Moltke had explained to me that 'Herr Doctor' did not like to talk about spiritual things during mealtimes. Rudolf Steiner mentioned some of the Baltic Keyserlingks and said: 'Yes, I know the Keyserlingks'. When I mentioned that I had once written to him in 1916, he said that he had not received the letter. After the table had been cleared I was to tell him about my life. I had had so many spiritual experiences which were inexplicable to me and about which I could not speak to others because of their lack of understanding, that I said to Rudolf Steiner that I would tell him everything without reserve, in spite of the fact that he might consider some of it to be crazy. But in his eyes it was not crazy, and by his interjections he often elucidated things, linking them to a deeper spiritual reality.

So I reported that my first memory was of watching the sun sinking into the Rhine at sunset: to my great fear and consternation this sun had pierced the depths of my soul, so that I could never forget it. And that happened to me when I was a little over one year old. 'Is it possible, Herr Doctor, for a child of one year to have such an experience and to retain a memory of it'? I asked, and was answered 'Yes, it is possible'. I did not recall anything of much importance about my childhood apart from this—except that, in the dusk of morning, I had seen unpleasant figures passing through the room in which there were closed coffins. But that ceased when we moved house. I remember then, when I was twelve years old, that I told another child that it was soon going to die—an occurrence which took place unexpectedly some weeks later.

The first really unpleasant thing which I experienced was when I bent over my little brother and discovered that the love within me turned into death-forces which threatened the child.

At that point Frau von Moltke intervened in an excited way: 'But, Herr Doctor, those are pathological thoughts'! Rudolf Steiner, however, answered quietly: 'No, what she says is right. In a former life the Countess experienced the fact that life on earth is death of the spirit'. I said to him: 'That happened to me again in the case of my own children—because of these forces they often drew near to the grave.' 'Certainly, that can be explained' said Rudolf Steiner, 'you saw your child in the spirit before birth and saw that you were giving it death and not life.' 'If I were to describe my earthly path up till the present,' I continued, 'I can only say it would be something I would not wish my enemies to have to undergo. And if I were to describe the effort it cost me to proceed along this road to the spirit, I could compare it with the picture of having to cross the globe barefoot with bleeding feet, and being told on arrival that I was to repeat the performance, not once, but more than once. In order to break through into the spiritual world I had to possess an amount of energy equal to that which would be required to walk round the world three times—abandoned to every danger and obstacle and only supported by my own spiritual forces.'

Rudolf Steiner had listened to me in silence, often nodding approval in his kindly way. So I continued: 'Often it seemed to me that it was not blood, but iron which coursed through my veins, urging me forward—yes, I had the feeling that my blood had congealed to iron in my desire to progress.'

That is where Rudolf Steiner interrupted to say: 'I can understand that very well. This iron comes from Mars.'

'I was confronted with cliffs,' I said, 'I penetrated them with my will, even though they threatened to fall on me and destroy me. It was immaterial to me—as long as, at the other side I could meet the Christ. I did not know about anthroposophy at that time, I only had my own will which worked like dynamite and would have blown everything to bits and urged me to penetrate right through the earth.' The look of approval from Rudolf Steiner encouraged me to proceed. 'This desire of mine to abandon the earth and search for the spiritual world started to appear very early in my life. I read Kant when I was sixteen and followed the thoughts of the various natural scientists and philosophers until I discovered that I could not get any further with thinking. There must be a gap in it somewhere and it did not lead me to any higher goal. Therefore I began to search elsewhere. My criterion was that with the Kantian type of thinking one could not grasp either the finite or the infinite. Thinking itself, however, demands that the one or the other must be valid. At this point my thought-power failed me and if I were to have penetrated further, I would have to have sought for a new direction of soul.'

I remember that Rudolf Steiner, sitting on the blue sofa in the corner, repeated thoughtfully the words: 'Yes, the finite and the infinite.'

'My study of the philosophers,' I said, 'was actually concluded around my 26th year. At that time, very slowly, new ways and inspirations from the spiritual world opened up to me, with their heights and depths. What I encountered there was not always very edifying—partly the most hideous things imaginable. I do not like to talk about it.'

As I had already spoken to Frau von Moltke about this, however, it drew from her the remark: 'I cannot understand how you could keep your sanity through all of it'! 'But,' I continued, 'I saw a previously unrecognised enchantment interweaving earthly events and I wanted to get hold of that. From it there issued forth words approximately as follows: *Though earth-life may undergo changes of mood, and destiny bring or take away from you what you hold dearest, look up to the power of love, of energy and goodness and there you will always find the same silent and blissful help for your soul.*

Rudolf Steiner looked at me with such benevolent and comprehending love that I decided to tell him about something very decisive which happened to me. 'In the winter of 1907 I suddenly became ill. I had an ectopic pregnancy. I was filled with thoughts of dying. I felt death approaching. It seemed like a call from the world of the stars which I would willingly have accepted, for I yearned for the light which shone behind earthly phenomena. There is always a parallel,' I said, as I turned questioningly towards Rudolf Steiner, 'for after this call from the light of the stars there followed a call from the depths.' 'Yes, certainly, there are always two' he replied. 'God gave me back my life, that is certain, for it was a hopeless case, the rupture had already occurred. I do not know if you are medically informed about the fact that in such a case one is beyond help?' 'But Herr Doctor knows everything!' broke in Frau von Moltke. Dr Steiner answered quite calmly: 'You are talking about a perforation of the tube. This being had already tried to incarnate into you once before. It was not a human being.' We were silent with amazement, then I continued: 'The doctors had left, and had replied to my husband when he had objected that they could not leave me lying there in such pain: "It is too late to operate and we have to save the morphine for the following days."

'When I heard that there was no more help to be had,

which I had already suspected, I then knew that the goal of my life was about to be attained, namely Christ. Either in death or in spiritual realms. White starlight enveloped me and I thought about the fever-racked son of the Captain of Capernaum. The thought passed through my mind: if I now see Christ, He will be able to give me back my life.

'Powers whirled around me and a Pillar of Fire appeared at my bedside. Such a mighty force streamed forth from it that I thought I would not be able to withstand it. It was like the tension of very high atmospheric pressure. I felt as though I would burst and I heard the words: "Depart hence, thy faith hath helped thee."

'Who was the Pillar of Fire?' I asked. 'That was Christ' replied Rudolf Steiner.

'There was a perfectly silent whiteness about me and out of this whiteness a sparkling life shot into me. The pains had ceased. The doctor arrived and assumed that this condition was the terminal phase. But life continued to ray forth; and when my husband asked what had happened I could only answer that he must turn to the account of the healing at Capernaum in the New Testament.

'From this moment onwards there was only one solution for us—I and my household were to serve the Lord. My illness was over and the whole of life took a new direction.

'Yet, as bright as had been the Light I had seen, so dark were the times I had to endure during the following years. It was as though I were passing through a living death. For years I experienced the gruesomeness of death in all its stages. The Light in all its inexpressible brightness had so dazzled my forehead that splinters of dead bone fell away from it: the Light of Christ conquered the bones which are the bearers of the forces of death.'

Rudolf Steiner asked me: 'Was that a spiritual event or did it take place physically?' 'It was physical,' I answered. 'The bones really were ejected. It was terrible. Everything festered and died off. The doctor extracted parts of the nasal bone through the gums with forceps and laid them out in front of me!

'I asked God in despair why I had to suffer all this and the answer was: "Because God can only cast anchor in the depths of a churned-up sea bed."' Again I met an approving glance from the eyes of Rudolf Steiner.

Then I related the approximate contents of my letter to him which had gone astray: 'I find it difficult to describe what I wanted to say at that time. I had come across the phrase in your writings: "The Midnight Call". I do not know what it means in that context, but it is the right expression for what I experienced at that time. At the sight of the primeval mother in the depths, the iron-strong will was released in a nameless pain. This sorrow was taken up in all its profundity by a burning love, as hot as the centre of the earth out of which we are born. This burning love was a promise of the transformation of pain into burgeoning life. So the will was changed into pain, the pain into love and the love was changed into life.'

Rudolf Steiner here interposed: 'Yes, the expression "The Midnight Call" is the right one.'

'Slowly I returned to the surface of life. I felt as if I had been submerged in a deep dark lake on the bed of which rested the primeval force. Then I stood in the midst of life once more.'

'I did not get the letter,' said Rudolf Steiner, 'I would otherwise still have it.'

'There is still something else I would like to tell you,' I then said. 'Powerful forces from the depths are rising up with violence, demanding entrance into my being.' Rudolf Steiner then replied: 'It is 1918 AD, so the powers coming in now originated in 1918 BC,' and he took a sheet of paper and drew the cosmic curve of world evolution, the nadir of which is the Event of Golgotha.

Then Her Excellency Frau von Moltke broke in, in her usual energetic manner: 'Now that this point has been settled, you should tell Doctor Steiner too how you passed through death!' But Rudolf Steiner interjected: 'No, she need not do that, I know already.' Deeply stirred by these matters, and by those which had remained unspoken, I blurted out: 'That, and many other things too, may be of great interest—but of what really stirs my soul I shall never be able to speak; perhaps you are able to see it Herr Doctor—but I can never speak about it'!

'Countess, you have told it to me already!' replied Rudolf Steiner, and Frau von Moltke interjected: 'But Herr Doctor, you have not spoken with the Countess before!' And I confirmed what she said: 'Certainly not, I have never spoken about it before!'

Rudolf Steiner said to me: 'Yes, you told me three days ago after the lecture.'

The disciple who was told 'I saw you under the fig-tree' must have felt something similar. To be received into the consciousness of this being was an overwhelming experience for me. Such a thing was quite new to me. We were dumb with shock.

Who was he? How could he know about something which had never passed my human lips? I was stunned that this person could look into my innermost being and that such a super-earthly event could occur between us. So I begged Her Excellency Frau von Moltke to leave me alone with Rudolf Steiner.

It is difficult for me to commit all this to paper. Many other things occurred during this hour together. Then we came to the end of our consultation, which was about the birth of the eternal Figure of Light in the reborn ether. I hurried to invite Frau von Moltke to come back into the room.

Rudolf Steiner then told me that everything had been correctly observed. 'I can only give you a few words of explanation today. Take up the following meditation into your thoughts:

In the evening:

In the beginning was the Word	
And the Word was with God	
And the Word was God	Be Thou in me
And may the Word be in me	(Before going to sleep)

In the morning:

And the Word was God
And the Word was with God   I am in Thee
In the beginning was the Word (after waking)

In the evening you must visualise the Midnight Sun of which you spoke, and again, in the morning, imagine the Spirit-Sun setting.'

When Rudolf Steiner noticed that I could not quite keep up with his dictation, he took the paper from me, made some corrections on it and finished writing it out. Then he made a few explanatory remarks in which Frau von Moltke joined and he asked me: 'Did you never write down what you told me?'

I answered: 'Yes, previously, but when I saw how thoroughly you have explained everything, I stopped doing so.'

Rudolf Steiner: 'You ought to write it down again, for it is always very interesting to see how it is experienced by each individual.'

Since then I have always written everything down.

Then Rudolf Steiner turned to me again: 'You must, of course, take what I now tell you in due modesty: you have the consciousness which will be customary in the third millennium.'

It then became clear to us why I was unable to understand myself. These are themes and problems which will only arise later, and find their realisation in humankind.

Rudolf Steiner then said: 'That cannot all be dealt with in a hurry. You must come to my apartment and we will discuss it further.' I looked at him in amazement: 'Do you have an apartment?' He answered me with a sweet roguish smile: 'Do you think I have nowhere to live?' And in his eyes were the words: 'Perhaps you think I am so spiritual that I can manage without one!'

Then he wrote down his address for me: Motzstraße 17, and told me the time of the appointment. In parting he said to me: 'You are in great danger from Lucifer, but the iron will protect you. You must not let yourself be influenced, however.' My reply was: 'Herr Doctor, before I came to you I had made up my mind that I would not even let myself be influenced by you!' A remarkably serious glance was the response to this remark.

Through conversations that I had with Frau von Moltke which often went on deep into the night, an intimate relationship grew up between us. I was accepted by her with an understanding belonging to a quite exceptionally deep feminine soul. The death of her husband, and the events surrounding him, the Chief of General Staff of an empire which had outlived its historical task, weighed on her. Germany was faced with the downfall of its monarchy, but in the summer of 1918 it still retained its impression of empire.

Frau von Moltke lived only a few minutes away from us. What a pleasure it was to enter her rooms in which everything was a reminder of her Swedish homeland. To that was added her Swedish-German accent which tripped with such charm across her lips. As I looked around, my gaze fell on reminders of Helmuth von Moltke, the victor of 1870. There was his death mask, his dagger, pictures of military postings.

After a short absence my husband returned to Berlin. During the war he had worked in the War Ministry. I told him, of course, about my friendship with Frau von Moltke and how I went to lectures with her and had even been invited to tea at her house with Rudolf Steiner. I said that he must get to know him and at all costs hear him lecture. 'Where are you carrying me off to this time?' he said with a smile, for he had so far always accompanied me for companionship's sake wherever my search had led me—also while I was with the Christian Scientists. And so he now went with me to visit Frau von Moltke, who told me afterwards that she would prefer to have Carl on his own next time. She had not met such a soul before, and then, searching for words, she said: 'He has a soul like velvet!' I was quite happy with this bond of friendship, and these two souls remained in unusually close connection to each other until the end of their lives. Their first meeting had been like a recognition, a mutual recognition.

And so Frau von Moltke asked for permission from Rudolf Steiner to bring Carl with her to the next members' meeting. Carl sat with us on the front seats and we had Rudolf Steiner's desk right in front of us. After the lecture we accompanied Frau von Moltke to her residence in the dark of night. Carl was uncommunicative and did not respond directly to our questions as to how he had liked the lecture. When we arrived home I asked him once more, but he again left the question unanswered. During supper I questioned him a third time and, as he again did not answer, I grew impatient and said: 'You must, after all, tell me your impressions of the lecture! You can tell me what you thought of it, even if you did not agree with it!' Then he broke his silence with the words: 'Well, if you must know: from now on, my whole life belongs to this man.'

He held to that in the truest sense of the word. Henceforth we could attend lectures regularly. 'Oh, do people not see who stands before them?' he once called out. He saw who it was straight away. And the trust with which Carl looked up to Rudolf Steiner was reciprocated. By the wish, and through the mediation of Frau von Moltke, Carl was then invited to meet Rudolf Steiner at his house. I do not know what was discussed there, but it must have been largely advice of a spiritual kind. Carl also discussed with him our private life in Silesia and the frontal sinusitis from which I suffered, and he also often visited him later. Since that time Carl always appeared to me to be deeply connected in soul to Rudolf Steiner for all time to come.

Memories of Eliza von Moltke

'Our Moltke' was a striking personality. She did not have much to say, but when she did speak her words had a solid, metallic sound. Her silences, her glance, her gestures had an air of majesty, but should one catch a glance of one of her rare smiles, one was charmed and enveloped by a motherly soul in whom primeval forces dwelt.

Rudolf Steiner, too, had an unusual connection to her. He was like someone known to her from the past. When they first became acquainted he was a young, unknown doctor of philosophy who taught in a socialistic Workers' Institute—and she belonged to the aristocracy. Yet soon he became a constant visitor at her house. For Frau von Moltke was one of the few people who recognised Rudolf Steiner at first sight.

The active, but at the same time, tactful way in which she introduced so many people, myself included, to Rudolf Steiner, must have been of great service to him.

Soon after I got to know her we spent many intimate hours together. Dressed in light-coloured clothing, and smoking—in spite of her needlework—she sat by the window and told stories about her life. She first made friends with Marie von Sivers, who was later to become Frau Dr Steiner. That was her introduction to the Theosophical Society and later to Rudolf Steiner himself. She was so attracted to his personality that without hesitation she acknowledged him, the much younger person, as her teacher. She discussed her private affairs with him and he helped her to judge the characters of her children too. On one occasion one of her daughters asked him in childish naïveté: 'Herr Doctor, have you really never danced?' whereupon she received the answer: 'Yes, your ladyship, in the roundelay of the planets!'

Rudolf Steiner also met the Chief of General Staff at the Moltke's house. I remember Eliza von Moltke saying that the latter had been very impressed by the wide range of Rudolf

Steiner's political knowledge. Frau von Moltke took the liveliest interest in everything which brought her into connection with the spiritual world. She was associated with Christian Science and came into touch with mediums. The interest in spiritualism, mediums and clairvoyance had a different nuance at that time than it does now. Eliza von Moltke followed up and tested everything—in spite of the fact that her husband, with whom she had an unusually firm and intimate relationship, treated all these things with the greatest scepticism. She even induced Rudolf Steiner to accompany her to such séances occasionally, as he had warned her against them. This certainly caused some remarkable situations to arise although, as Frau von Moltke reported, Rudolf Steiner remained a completely passive member of the audience.

The famous uncle, the Field Marshall, who played such a decisive role in the fate of Germany in 1870, lived in the same building as his nephew, at the Headquarters of the General Staff in Berlin until his death in 1891, so that Frau von Moltke was able to look after him. She told me very movingly about his death: before retiring to bed the elderly gentleman usually had a game of scat. On the last evening, as he was on the point of retiring, he looked so pale and delicate that she was concerned for him, and remained for a while at his bedside. He dropped off to sleep, and quite unnoticed, his soul slipped into the other world.

The following morning the log-book of the previous day was presented as usual to the nephew, according to the rule. In it was noted by the two sentries who kept watch at the front entrance, that His Excellency the Field Marshall, had passed through the main gates at midnight and walked towards the Spree Bridge by himself. It was conspicuous that he was without either his dagger or his cloak. In addition to that, two young officers who happened to be passing the main gates at the time, had seen him emerging from the front door. They thought of speaking to him, but got the impression that the elderly gentleman would perhaps rather be left undisturbed to take a breath of fresh air. Out of a certain feeling of concern they followed behind him, but at the Spree Bridge he suddenly vanished from their sight.

It was only on the following morning that they heard about his death. They were so impressed by this event that they published it in the newspaper. Frau von Moltke, too, wanted these facts to be remembered.

When we travelled to Dornach together in 1920, I was allowed to sit with her on the front row next to Herr and Frau Dr Steiner during all the performances, and had an opportunity to listen to many of Rudolf Steiner's remarks. I received a sharp rebuke from Frau von Moltke on one occasion when a speaker who overstepped his time to such an extent that the public began to laugh loudly at him, was interrupted by Rudolf Steiner and asked to draw his talk to an end. As the man stepped down from the platform he gave his hand to Rudolf Steiner, but the latter only inclined his head politely. In order to alleviate his obvious embarrassment I offered him my hand, at which Frau von Moltke glared at me: 'How can you accept a person whom Rudolf Steiner has refused!' Yet Rudolf Steiner gave the suggestion of a smile and I imagine he approved of it.

During question time a young man asked Rudolf Steiner about the problem of marriage. Rudolf Steiner answered, in serious vein, that it was a social institution; and then said quietly and jokingly to his wife: 'He probably wanted to know from me whether or not he should marry.' On another occasion two speakers had got into an argument with one another. At that Rudolf Steiner said to Her Excellency: 'When mice eat each other, only their tails are left!'

Rudolf Steiner had complete trust in Eliza von Moltke, just as she had towards him. Their relationship to each other, which was one of mutual love, friendship and spiritual understanding, remains as a clear picture in my mind.

In the same way the loving friendship which she immediately felt towards my husband is only to be explained in terms of a karmic connection extending over many lives. Once, when she was in great sorrow while staying with us in Koberwitz—Carl was absent at the time and was not expected to return very soon—he unexpectedly came into the room. She stood up immediately, stretched out both hands to him and, with hardly-restrained tears in her eyes, said to him: 'Now everything will be all right again!'

One morning—contrary to her usual practice—she got up at 7 am in order to have breakfast with Carl. After he had driven off to the Ministry she was standing by herself in the room when I entered and she said happily to me: 'To have met such a man has made life worthwhile!'

Occasionally she talked about her meetings and conversations with Rudolf Steiner. Her exuberant expression as she related the following incident remains vividly in my memory: Two theosophists had taken their leave of Rudolf Steiner in Munich as they were about to leave for India to look for a great initiate. They had discovered from Rosicrucian sources that this initiate was alive today and had been born in Hungary. The smile with which she said to me: 'They could have spared themselves the trouble!' can easily be imagined.

She also told me about something which had occupied her mind time and time again: in speaking about a manuscript in the Vatican Library dealing with the dominion of Lucifer, Rudolf Steiner had said: 'A copy of this is owned by a personality who is seriously misunderstood by the world at large, but is now becoming of interest to those engaged in historical research—I could also say was owned—but that could only bring about unclarity—therefore I say: the Count of St. Germain owns such a copy.' As Frau von Moltke was such a lucid spirit who never jumped to premature conclusions, this formulation gave her much to think about.

In Berlin she told me as we crossed the street at the Potsdamer Place, that she had been here with Rudolf Steiner before the First World War. He had said to her then: 'Look at these stones, everywhere grass will be springing up between them and the houses which stand here will be in ruins.' On another occasion, when they saw an acquaintance driving past in a carriage, Rudolf Steiner said: 'That man has hardly any strength of ego to see him through till the end of his life, let alone to suffice for a new incarnation.' Because of that the word went around that there are egos which are becoming extinguished at the present day.

Eliza von Moltke was a person of many parts. She was both energetic and practical and at the same time she had a spiritual awareness and sensitivity which made her conscious of the seeds of destiny which had placed her and her husband at a dramatic crossroads of German history. She immediately grasped the importance of the 'threefolding of the social structure' and made it possible for Rudolf Steiner to have talks with leading personalities.

She took an active part in activities centred round her husband in the General Staff and in the immediate presence of the Kaiser. And when events leading to the declaration of war with Russia came to a dramatic head, both Moltkes were aware of the fact that self-sacrifice was demanded of them.

What Frau von Moltke told me after the war painted the catastrophic situation in much sharper outline than did the publication of the Moltke letters.

Yet the way she supported her husband through the most oppressive situations, and after his sudden death met the hostilities which were intended to blur the real issues, was of such magnitude and entailed such inner truthfulness, that it evoked the deepest sympathy and highest regard from all who knew her.

Eliza von Moltke remains firmly connected with all those who are concerned with Koberwitz and the Agricultural Course.

Encounters 1919–1921

A year had passed since the memorable hours I spent in Berlin with Rudolf Steiner, nor had I seen him since. But with the whole might of my soul I was determined to find out the spiritual meaning of the words which I was privileged to receive during those six unforgettable afternoons. Then, in September, a letter arrived from Frau von Moltke to say that Rudolf Steiner was expected to give lectures in Berlin, and she urged me to come there.

It is true, I had not seen Rudolf Steiner in the meantime, yet not a day had passed without me seeking him in spirit and turning my gaze upon this magical figure who had been sent to us like a messenger of the gods. Mighty and ever mightier this figure became known to me as though surrounded by divine radiance. And now I was to see him once again. Part of me was full of trepidation—what would be my experience? I would see before me a human figure, whose essential being I took to be capable of rising consciously into the highest hierarchies. How would it be if, in my loneliness, I had only conjured up a Luciferic image which would disintegrate once I stood before him again? How could I bear that? That was what was worrying me as I got ready to go.

The train arrived late in Berlin. I should have arrived punctually for the lecture, but I noticed to my dismay that I had been given a wrong house number in Potsdam Street. I wandered about looking for the right place, and fearful of not being allowed into the lecture; yet my anxiety about what I should meet there was still greater. Would the dream I had indulged in for a year vanish away? Would I find before me an ordinary human being like everyone else?

At last I found the correct house number. I listened at the door to the voice of the speaker, whose lecture had already begun. It didn't sound any different from the voice of any other speaker. I stepped quietly inside and found a place on the back row. And something then took place which belongs to one of the most powerful events of my life: Rudolf Steiner looked up and his glance penetrated across all the other people to the back row where I was. My desperate thoughts had been perceived by the ears of his spirit and his eyes said to me: 'I am much more powerful in the spiritual world than you have hitherto thought possible!'

Heavenly spheres opened behind him. I gazed down into the golden fiery background which encompassed the depths of his soul. The force proceeding from this sight was hardly bearable. And ever and again, Rudolf Steiner looked at me anew with the same strong gaze. Now I knew that I had been living in reality and not in illusion.

As though stunned I walked back to my flat at night through the busy streets. I could still see the fire-golden, subterranean depths into which Rudolf Steiner allowed me to look as into a part of his being.

I experienced the divine ground of the world as though through an initiation. During the whole night I was wrapped in the fire-spell of divine powers.

Soon after this, however, I had the chance to see Rudolf Steiner from a quite different angle. It was when he was involved in publicly promoting the Threefold Social Movement. He had just returned from important political negotiations in Stuttgart, in which he had won the broad agreement of the general public, but had failed to convince rigid bureaucrats and overcome the fear of Trades Union leaders. The war had ended a year previously and owing to the revolution all free political life had been forcibly suppressed. Rudolf Steiner had been asked to lecture in Dresden, but on the eve of departure he was informed about restrictions imposed by the socialist town council. He was only to be allowed into Dresden on condition that he followed those instructions. With righteous indignation he snapped at those who had delivered this message: How could they expect him to undertake the journey at all under such circumstances? It was truly a godlike anger which was fired against these folk!

Rudolf Steiner foresaw a force of complete spiritual vacuity approaching, which would draw Europe into an insoluble chaos. People, he felt, were convinced by the newly founded, threefold social organisation, to which they looked for help; and yet now the possibility of protecting Germany from endless misery would fail through mere narrow-mindedness and fear.

I looked at Rudolf Steiner's countenance—it was white with anger at the human beings among whom he had to live. All the people in the room were dismayed, for he had thus announced the end of the Threefold Social Movement.

The lectures to the Schopenhauer Society in Dresden were the scene of much turbulence from the very start. We decided to accompany Rudolf Steiner. Frau von Moltke set about the preparations with such vigour that it made her look quite young. On the morning of departure her son Bill had to go on ahead to the station to book a compartment for Herr and Frau Dr Steiner, Mita Waller, Frau von Moltke and myself. We already spied Rudolf Steiner in the press of folk on the platform, his heavy briefcase, with his papers, hanging from his shoulder. He entered the compartment with the porter, who swung the case vigorously onto the luggage rack above me. 'Stop, stop! don't kill the lady,' said Rudolf Steiner to calm the man's vigorous efforts. Then someone poked his head in: 'Is there a place here?' 'Yes, there's a place free,' said Rudolf Steiner and pointed to a seat strewn with our belongings. How conscientious was Rudolf Steiner even in everyday things! Then we drank some mineral water—all out of one glass. Rudolf Steiner spent most of the journey immersed in his newspapers and brochures.

In Dresden we were treated like the retinue of a VIP. During the lecture at the Schopenhauer Society I again sat beside Frau von Moltke, who was next to Herr and Frau Dr Steiner, so that I got a clear view of the proceedings.

Rudolf Steiner mounted the dais and delivered the lecture he had been invited to give. He spoke appreciatively about Schopenhaur, but not in the way the pupils of this philosopher were accustomed to. When he had ended, and before he could leave the speaker's desk, a gentleman belonging to the council of the Schopenhauer Society sprang onto the platform and pulled the lecture to pieces with patronising disdain. He ended his speech in an abusive manner, saying to Rudolf Steiner: 'No one would believe that a genius who proclaims new wisdom was speaking to us here!'

I could hardly sit still for growing fury, and a wild energy was spreading over Frau von Moltke's features. How would Rudolf Steiner react? I looked across at him but was unable to detect any change in his countenance. He was able to retain his dignity because what had been said could not lessen the truth of his knowledge. The audience dispersed, but in the passageways people stood in excited groups. I approached the lively, diminutive Frau Waller. 'I could box his ears till he lost all his senses!' she called out.

On the following morning I accompanied Frau von Moltke to a eurythmy practice. Only a few people were present. Marie Steiner was reciting. Rudolf Steiner drew near to me and imitated the rolling of her Baltic 'R' in such a droll fashion and with such a comical expression as if to say that the happenings of the previous evening had not affected him in the least and that I could go on laughing if I liked.

Next day, when Rudolf Steiner gave a lecture to members, the lady who had invited him to come to Dresden stepped onto the platform and thanked him in the name of all those present. Then she pulled out a note and, with joyful satisfaction in her features read the following words by Christian Morgenstern, who was still relatively unknown at that time:

To Rudolf Steiner

> As on a rainy day one can forget the Sun,
> Which shines on brightly still without a pause,
> So, on a dreary day, we might forget
> Your image, until, heart-shaken, ever and again
> And blinded by the light, we find once more
> Your radiant Sun-Spirit inexhaustibly
> Shining towards us jaded wanderers, on and on.

The daylight hours in Dresden were taken up with discussions, but two evenings were set aside for theatre trips. We had booked seats for *Iphigenie,* and while we were handing in our things to the cloakroom attendant I could observe again how Rudolf Steiner was able to know what others were thinking, just as people otherwise understand the spoken word. I did not feel worthy to regard myself as part of Rudolf Steiner's cortège and had put my coat a little further away from the others. Then Rudolf Steiner approached me and, without saying a word, took my coat and gave it to the attendant along with his own. When we left the theatre he helped me into it and enquired in a professional manner if the collar should be inside or outside.

On the following evening we had tickets for *Schneider Wibbel.* Suddenly Rudolf Steiner attracted my attention: 'Look how that child over there is laughing. This play is a lot of nonsense—but laughing is healthy.'

The next morning I had to go to Rudolf Steiner to speak about a rather unpleasant situation with regard to my father. I had written everything down and gave him the note with the remark that what I had written was not very nice . 'Oh, is your writing so bad?' he said laughingly.

At this time Rudolf Steiner was looking for land to add to the *Kommenden Tag* estates which were administered by Carl and Emanuel Vögele. I did not know at that time how tight the threads had already been drawn which were to lead us to Koberwitz. Rudolf Steiner asked me, as he pulled out the *Schlesische Zeitung,* who had sent in the article about agriculture, and I replied that it was I who had done so.

The thing which strikes me as important from those Dresden days is the following. As one may well imagine we were often taken up with the question: 'Who is this mysterious Rudolf Steiner?' Frau von Moltke knew that I was clairvoyant and that Rudolf Steiner had told this to Marie Steiner and Fräulein Waller, and so she could not avoid asking my opinion about this. I said: 'Rudolf Steiner's outer form appears to me to be a covering, behind which a spirit-form of shining gold rises up.' Finally I formulated it in the following way: 'He is the bearer of a power which Christ designates as "the Comforter", which He would send us to guide us to all Truth.' Frau von Moltke did not reject this thought, but replied that she could not accept it, because Rudolf Steiner was decidedly opposed to any deification of his person by his pupils. He would definitely reject that. So I asked her to ask Rudolf Steiner herself, in some way or other. Now, she had been called to speak with him and was determined to bring up this question. I waited for her in the hotel and can still see how she sank into the armchair and said: 'Yes, he confirmed what you say.'

Next there follows a memorable discussion.

I felt myself especially favoured by destiny in those days as I made my way through the streets to Rudolf Steiner. I was sorry for those who could not have a similar experience. Was there a single one among those I met who was as fortunate as I? I rang the bell punctually at the house where the Steiners lived. Marie Steiner and Mita Waller met me at the door as they were setting off for a eurythmy practice at the Opera House.

These couple of hours I had alone with Rudolf Steiner cannot easily be described, for this teacher, who also taught through his spirit-body, often turned spoken words into a spiritual event.

After one such conversation Rudolf Steiner asked me about Carl, for I had already told him in Berlin that my father was threatening to dismiss him every time he took an initiative of his own. Today, however, I was able to report that, as if by a miracle, we had unexpectedly received a large sum of money from security bonds which at least made us financially independent.

Only then did Rudolf Steiner open what I had written for him describing our difficulties. The trouble was that Carl had to run the estates of our family's company from our city residence. 'Shall we give in, and renounce living in the country?' I said.

Rudolf Steiner: 'Why did your husband never say anything to me about it, when he visited me in Berlin?'

I: 'Men often do not talk about such things. My husband

thinks that we have fought so long for a house in the country and that it is not God's will, so we should be resigned to it.'

Rudolf Steiner: 'Perhaps your father might also die someday?'

I: 'No, Herr Doctor, he won't die. He will continue to plague us for decades.'

Rudolf Steiner: 'No, he won't die immediately.'

I: 'And perhaps Carl also thinks that we should treat the old man gently.'

Rudolf Steiner (with energy): 'Not so! Look at it this way: when one crosses a person in this life they get angry, of course—but then, when they come into the spiritual world, they are grateful for it!'

In the course of the conversation I asked how it was that such wishes remained quite unfulfilled, whereas in other cases I had always found help, even in the minutest detail. When I lacked a piece of material to repair a dress, an inner voice told me the shop where I could buy it. Rudolf Steiner rejoined: 'In the case of a piece of material that works quite all right, but when it is a matter of land, quite different laws are involved. I will write to your husband about it.' Next evening Rudolf Steiner gave me a letter which said that in regard to a person who had been thwarted in a good intention, the spiritual world feels thwarted too and expects the person to work for his original goal in spite of the hindrances (see letter p. 222).

Then I drew a plan of the estate grounds which were thought to be a possible site for us to live. Our fields bordered the southern boundary of the city of Breslau, immediately in front of which was the sugar factory. I told him about Lohe, the mysterious old moated castle with the fortified tower which I would so like to have had as my residence. I had already drawn up a plan with an architect for the alterations, and had got so much satisfaction out of it that something was simultaneously set free in me—so that now I did not need to own it any more.

Then we spoke about Koberwitz. Rudolf Steiner would not

cease speaking about it, in spite of my arguments to the contrary, and finished by saying: 'Ask to have Koberwitz!' I told him that, in order to make it impossible, my father had concluded a six-year agreement with my elder brother for him to live there. 'Ask for Koberwitz' insisted Rudolf Steiner. 'Your father will find another house for your brother in place of it! It might, of course, result in your brother no longer wishing to know you!'

One must imagine how the words: 'Ask for Koberwitz' brought a quite new direction into our lives. I suppose it was 15 years that we had been fighting for a place to live in the country. My father was afraid that he might lose his authority over the estate if Carl, its manager, were now to live there among his people. My father was a despot who had built up a small kingdom with his 30,000 acres of land. Although he had entrusted Carl with the administration of the land, he was very conscious of the difference in outlook that his son-in-law had. As soon as Carl had come up with this request he had threatened to dismiss him.

But now, since Rudolf Steiner had sent him this letter, Carl went one morning to my father's office and broached the subject once more with calmness and confidence, saying that it was a waste of time and strength for him to have to travel between the town and the country and that the enterprise would be easier to manage if he lived on the estate.

He returned home moved and shaken: a miracle had happened. My father was like a changed man and, remarking that he needed more information about it, had been quite approachable.

Then everything happened all at once. At the next council meeting my father proposed the alterations to the Koberwitz mansion, which were approved without more ado. Yet even more agreeable was the fact that we were aware of Rudolf Steiner's protecting forces surrounding us. Just as he predicted my brother was given another mansion. And one evening as I was crossing the courtyard at Koberwitz, I encountered my brother, who passed me by without greeting me.

Only now, after many years, it has become clear to me what Rudolf Steiner achieved when he helped us to overcome the difficulties. For what had seemed impossible came about: my father, this despot, gave in! Routed were the demons who had wanted to prevent the estate being transformed into consecrated ground! Now I know it! Rudolf Steiner possessed the power to dissolve wickedness and transform it into good!

And then came the joyous time in which, together with a skilled architect, I was able to convert the old mansion into a bright and pleasant dwelling place—as beautiful as one could wish to have as a country residence, with a park, gardens and a lake.

The move took place in the early spring of 1920—it was a festive occasion. I should like to perpetuate its memory by a description of the house to which Rudolf Steiner, by his presence, brought so much happiness and hope for the future:

After the drive had been newly designed and the paths and the lake had been cleaned up, a log-cabin was built on the shore of the latter and a rowing-boat lay alongside. We arranged a guest-room on the assumption that Rudolf Steiner would soon come to stay with us. Frau von Moltke was our first guest.

The whole house had about sixty rooms. A left and a right wing were built onto the back of the middle section, which had a terrace overlooking the front drive, so that a herbaceous border could be introduced in all its blossoming, colourful glory. I arranged for the letters PSSR (per spiritum sanctum reviviscimus) to be carved over the smaller side door—people thought they were the initials of the building firm and took no further notice of them.

One entered the south-western part of the left wing, which we had arranged as our own quarters, through a small domed vestibule, the curved walls of which were lined with blue Gobelin tapestry. Elk and deer antlers were hung there. Then one entered a cosy hall, crossed some black and white marble tiles and climbed a black-oak winding staircase to the first floor, past a huge elk which I had shot in Norway. A large painted-glass window hung with red velvet curtains, an old grandfather clock from Denmark and a carved ship's model, which hung over the great round table in the corner, made the hall very snug. Double doors opened into the large dining room which was part of the central block, so that at mealtimes one could always look out over the herbaceous garden. On festive occasions, as when lectures by Rudolf Steiner took place, or at Christmas or Whitsuntide, the folding doors were thrown back, so that the hall and the dining room became one large room. On the ground floor there was an adjoining kitchen and workroom to gladden the heart of every countrywoman, which was also closely inspected by Rudolf Steiner.

Ascending from the hall one was received into the music room. Its walls, decorated in warm violet, were divided up by large, gold-framed areas spanned by Parisian tapestries in the style of Louis XIV, in which colourful red parrots were depicted, swinging among tender pale lilac blossoms. The red silk cover over the grand piano and the white Empire-style furniture gave a pleasant air to the whole room. There was a harmonium and all kinds of other instruments in it. Carl loved to play the violin and often invited other music lovers to play quartets or trios with him. I can still visualise Dr Schwebsch here, playing Wagner with enthusiasm, or the Viennese violin-maker Thomastik, who demonstrated his instruments for us.

One next entered the study through wide glass doors. Dark furniture stood out against light grey wallpaper. A Persian carpet covered the floor of the more than six-metre-long room. Here stood Carl's writing desk; and a huge eagle, the work of a wood carver from Berchtesgaden, stood on a tall pedestal and dominated the room. The view from the large windows of a wide recess looked out over the park. This is where Rudolf Steiner received those who came to him for advice every day. Carl preferred to work in a smaller room next to his bedroom.

From the music room one came into our completely

secluded small flat, which we put at Rudolf Steiner's disposal when he visited, so that he could remain undisturbed. In it was my sitting room with deep blue walls, light green furniture and a green carpet.

My sons' rooms and the guest room were on the upper floor. Every window in the house looked out into nature. The whole surroundings were very pleasing.

Count Carl von Keyserlingk

If we look for the spiritual currents to which Rudolf Steiner linked his new impulses, we would find many individuals who brought this connection through their personal destiny. Carl Keyserlingk was also one of these. He brought with him a European, rather than a German destiny.

My father-in-law, Eugen, was a Balt, descendant of the German knights who had farmed the land in Eastern Europe, where education was valued in a way hardly rivalled by any other country. Handwritten manuscripts of Kant were preserved at his ancestral home at Rautenburg am Haff, where he had been a teacher. The Balts were just as much at home in Petersburg and Moscow as they were in Berlin and Paris. They surpassed one another, not so much in luxury as in having world-wide interests, entertaining important guests and having highly refined domestic arrangements.

My husband's mother was the younger daughter of the Bavarian Ambassador von Dönniges, who, as successor to Humboldt, was called to Rome and was a frequent guest at the table of Louis II, the friend of Richard Wagner. She was not such a renowned beauty as her sister, the famous Helene Rakowitza, who was the talk of the town because of the duel between Lasalle and Count Rakovitz, but she stood out on account of her writings; and her connection with the nobility of Europe was decisive for the childhood of her sons. So Carl spent some of his childhood in Rome and may, without conscious awareness of it, have heard the name of Rudolf Steiner, who held his first Munich lectures at the hospitable home of the Dönniges.

Carl's father had settled in Silesia. He was a naturalist and wanted to live closer to Central Europe's furious pace of developments in Natural Science. His collections were later bought by the Natural History Museum in South Kensington.

Carl was born on 14 August 1868 in Jakobsdorf, on his

father's estate at the foot of the Landskrone. He became an officer. As he had been brought up in the country and had enjoyed a thorough agricultural training with a view to taking on the care of the estate later on, it was an obvious choice for him to be put in charge of the goods stores in my father's sugar factory after we were married. That was why he was appointed by the War Ministry to supervise the distribution of foodstuffs in Budapest during the First World War and was later called to Berlin. The change which took place in his life as a result of meeting with Rudolf Steiner is the subject of these recollections.

The qualities that had been preserved for centuries in the German Chivalric families of the East—loyalty, steadfastness, a living connection to the soil and a free, world-embracing spirituality—were those that Rudolf Steiner wished to connect with for the Agricultural Course, even exoterically, through people like Carl Keyserlingk.

Carl's first meeting with Rudolf Steiner has already been described by me. He was then constantly in the presence of Rudolf Steiner, who gave him the directorship of some of the smaller estates in Württemberg when the *Kommenden Tag** with its business undertakings was founded.

The fact that Carl had also taken on work in Württemberg aroused an unpleasant stir in our sugar factory. There had been difficulties between Carl and my father for a long time, because they were of very different dispositions and my husband wanted to introduce independence into the social relationships in the firm and in the administration of goods, which my father would not allow. Carl's conversion to anthroposophy was therefore a welcome opportunity for my father to turn the directors and trustees against him. 'Departmental independence' was the advice which Rudolf Steiner always advocated when Carl spoke to him about these difficulties. But that is just what my father wanted to prevent. There was a danger that the initiative of the younger ones might outstrip that of the older generation.

---

* A company set up in accordance with threefold social ideas.

The plea which Carl had been making for years to be given one of the estates as his living quarters had been flatly refused by my father. I have already told you how this came about, after all, through the help of Rudolf Steiner. But this would still not have resulted in the Agricultural Course being given there if it had not been for the fact that Rudolf Steiner recognised in Carl the steadfast trust and loyalty needed by people who want to cultivate the earth and take responsibility for the nourishment of plants and animals. I will recount two episodes which will make this clear.

In the winter of 1919/20 Carl had sought Rudolf Steiner's advice in Stuttgart prior to an important trustees' meeting in Berlin, in which he wanted to push through his idea of 'departmental independence'. He said goodbye to Rudolf Steiner after his interview and was surprised to receive a note from him with the remark: 'Here is your appointment as manager of the *Kommenden Tag* estates.' Somewhat surprised, as nothing concrete had been said about it, Carl put the note in his briefcase and set off for Berlin. There the mood was extremely tense. But, nevertheless, Carl demanded complete autonomy over the property, according to the advice he had received from Rudolf Steiner. The trustees withdrew to consider the matter and then Carl was told that his application had been rejected and that it was left to his discretion to decide if he would resign his work in Silesia. Then it suddenly struck him! The new appointment! Doctor Steiner has safeguarded me! And he drew out the appointment paper with the remark: 'As this outcome was expected, another offer has been accepted.'

The gentlemen were so taken aback that Carl's appointment was renewed and finally agreed upon.

But my father did not give him any peace. A quarrel then arose between them, which ended in such a way that Carl was to resign from the sugar factory in Klettendorf for half a year, until 1 October 1920 initially.

Carl loved his work and was likewise loved and respected by his colleagues. This turn of events had therefore shaken him very badly. He, however, only gave expression to it with the words: 'Now my life is free to serve Doctor Steiner!' Then he took up his violin and played. Soon afterwards he travelled to Stuttgart to the *Kommenden Tag*. In October the final decision would be made.

At the end of September, a few days before he was due to return to Silesia, a letter came to him from my father which said that the best thing would be for him not to return to work in Klettendorf—he was asked not to be seen there again. One morning my husband took the letter with him to the *Kommenden Tag* office in Stuttgart, where he was awaited by Rudolf Steiner, who had been there for some days. When Rudolf Steiner had read the letter in his usual careful way, he remarked: 'Well, the letter is not exactly polite—but you should go back there nevertheless!' My husband was very surprised, because this piece of advice was actually an impossibility. Yet Carl accepted it with inner confidence. He tried once again to explain the situation to Rudolf Steiner, but the latter parted from him with the brief and evenly pronounced statement: 'No, return without hesitation!' Carl could not even ask to be provided with a detailed contract. He had lost his job and had not been given a new one—it was, at the very least, a most unusual situation. But his trust in Rudolf Steiner was so great that he departed for Silesia the next day. There he was confronted with a situation which he had not at all expected. His wages and upkeep had been stopped. His horses and his coachman had been sent to another estate. His chauffeur had been transferred elsewhere and the garage stood empty.

In this situation Carl obeyed the voice that had given him advice. He travelled by train next morning to Klettendorf, walked to the factory and sat in his office. No one greeted him, nobody brought him the letters or called in at the door. The employees seemed to have been given orders to this effect. When it was time to leave, he went back home. The same thing happened next day—he travelled to Klettendorf, spent the day quietly at his desk and on the third day it was repeated once more. But then suddenly there was a knock at the door. The head of the factory stood there, white in the face and stuttering: 'Your Honour! A deputation of the workers has arrived—a general strike has been declared at the Trades Union Headquarters in Breslau if you do not immediately take up your post again!' Carl was very moved by what had taken place between him and Rudolf Steiner, which cannot be described in words.

The results of this situation exceeded all expectations. Not only was his authority restored to him, but a new contract with complete freedom to work as he thought best and a high wage was given him—the unlimited power of my father had come to an end.

Happy employment, carefree enjoyment such as we had never previously experienced now began, and reached its peak when Rudolf Steiner came to stay with us at Koberwitz on three occasions.

Rudolf Steiner's first visit to Koberwitz 1922

30 January–2 February

We sat at table in a light-hearted mood in Koberwitz with our fellow guests, in particular the priests Rudolf von Koschützki and Rudolf Meyer, who lived with us permanently. Carl had just got back from Stuttgart and called out: 'Just imagine, Doctor Steiner is coming to us here!' Amazement and questioning began and Carl reported that Rudolf Steiner had told him that he would be making a lecture tour at the end of January and would also be visiting Breslau. When Carl discreetly enquired if he would do him the favour to stay with us in Koberwitz Doctor Steiner accepted in the friendliest manner. Soon appeared more detailed information in the beautifully clear handwriting of Marie Steiner, to say that they would come to stay with us with their foster-daughter Miss Waller for about three days, during which time the public lecture was to be given in Breslau.

The members there were competing to offer hospitality to the Dornach eurythmists. A booking for the large concert hall had been made for the occasion with Herr Wolf, the director of concerts. When we asked Herr Wolf if he knew Rudolf Steiner, he said that he did not know him personally, but he knew that a public lecture by him had been fully booked weeks ago already.

At last the telegram arrived announcing Rudolf Steiner's arrival on 30 January. Marie Steiner was to come directly to us, whilst Doctor Steiner himself was first to attend to the business arrangements with the concert directorship. Thus I travelled later with Marie Steiner from Koberwitz directly to the theatre where the dress rehearsal for the eurythmy performance was to take place. I was sent by her from there to collect Rudolf Steiner, as she had some questions to ask him.

So I drove to the Savoy Hotel, where I found Carl and Rudolf Steiner drinking coffee together.

One can imagine that it is not easy to approach Rudolf Steiner, especially after not having seen him for some time. One experiences a sense of awe, mixed with a feeling of uncertainty, like that of a schoolgirl called up before her headteacher. Rudolf Steiner always helped to ease such embarrassment, which he immediately recognised, by his skill and good-heartedness.

So it was that we drove to the theatre and then, after Rudolf Steiner had given his instructions at the rehearsal, we drove back to Koberwitz for the evening meal.

Everyone was happy at being allowed to meet Rudolf Steiner in such a small circle of people. When I said to him: 'Haven't we made it nice here?' an understanding laughter shone in his eyes.

After the tiring journey all of us wanted to retire early. Our young farmers, such as Erhard Bartsch and Imanuel Vögele, rose to go. Rudolf Steiner held out his hand to them all with a friendly gesture. The old butler bowed in silence, but to his astonishment Rudolf Steiner held out his hand to him in the same friendly fashion to wish him goodnight.

Perhaps one cannot imagine today how such a simple gesture as this did not belong to the conventions at that time and would not be expected by anyone. But Rudolf Steiner looked further ahead into the future. Then I introduced Fräulein Pietsch to him, who had been with us for thirty years, but he said in a kindly way: 'That is not necessary, I know her already.' Fräulein Pietsch however declared: 'Oh no, Herr Doctor, I have never seen you before.' But he said to her again: 'Yes, it is so, I already know you!' And so all felt at home with him from the very first evening.

All had breakfast in their own rooms in the morning. Then Carl's brother and his wife came to hear the evening lecture. We noted with some anxiety the ever increasing snowfall, for it had been announced that all the railways in Silesia had joined the strike, and on the next day but one Rudolf Steiner was due to give a public lecture in Stuttgart, and Marie Steiner a grand eurythmy performance in Prague.

A great wish of mine now found fulfilment, for Rudolf Steiner asked me if there was anything new that I had been working at, and said that I should come to him after breakfast. Thus I was able to speak with him for two hours about matters to do with spiritual science, particularly things to do with astronomy—of the place of the earth within the cosmos. After we had discussed this I asked him: 'Am I allowed to speak to other people about the things which I experience?' 'Yes,' he answered, 'of course you should do so! People would receive great strength from that—only you must be careful to whom you tell these things, so that they are received with due respect and not made ridiculous!'

I: 'Frau von Moltke does not believe me either.'

Rudolf Steiner: 'So you quarrel about it do you?'

I: 'We will get over it again all right.'

Rudolf Steiner: 'Yes, you must keep in touch with one another.'

I: 'I still want to ask you, Herr Doctor, why I do not receive guidance any more. Formerly I did not have to bother about anything; all was arranged for me—I even thought that the train was held up so that I could catch it—now it sets off without me.'

Rudolf Steiner: 'That was bound to cease sometime—you have to think for yourself now! Actually you stopped being directed the moment you came to me.' And with an infinitely kind expression he added: 'You will perceive that you are nevertheless being guided.'

I: 'If I am not guided any more now—how can I act? Is there such a thing as instinctive knowledge? That is what guides my husband.'

Rudolf Steiner: 'Oh yes, it is all right for your husband to follow his instincts!' And with a spark of mischief in his eyes: 'He is able to do that because he is completely unselfish.'

He held back what he was thinking—that I was not so. Then there was a knock and Rudolf Steiner was called away. He stood up and said to me: 'You are not angry that I have to go?' 'Of course not,' I said. And again, when he was halfway to the door he turned and said: 'You are sure you are not angry with me for leaving you?' I: 'No, of course not, I am happy to wait!' But at the door he turned once more with the same question, so that I could only laugh. He was referring not just to the present moment, but to the fact that I now had to become independent.

At lunchtime topical political events were discussed. My brother-in-law was engaged in politics and so the rail strike provided the occasion to discuss the journey from all aspects. Carl had telephoned around and had succeeded in getting friends who would try to fight their way in stages by car to Stuttgart via Liegnitz, Dresden and Nuremberg. It continued to snow without a break.

At seven o'clock the great lecture began in the concert hall in Breslau, which must have been a great disappointment for many who were there. Rudolf Steiner spoke in such a dry and scientific way that only a thinker grounded in philosophy could follow, whereas people had expected something sensational. But Rudolf Steiner knew what he wanted to achieve. He said afterwards that one could talk an audience out of his lectures just as well as talk them in. This overpowering stream of people was not what he was aiming at.

The following morning revealed Rudolf Steiner to me in a new light. He was standing in conversation with Carl and his brother, the regional president of East Prussia, who held an influential position. The latter was interested in the Threefold Movement and seemed delighted to become acquainted with an informed and competent diplomat in Rudolf Steiner, instead of the starry-eyed idealist he had expected. Rudolf Steiner was also able, through some of his remarks and a winning smile, to completely convert my sister-in-law, who up till then had behaved very sceptically towards him. I think there was no one who could remain unaffected by the radiating goodness of Rudolf Steiner. I heard that Kühlman was being discussed. This interested my brother-in-law very much because, as a politician, he saw immediately that if the advice that Rudolf Steiner gave had been followed, the end of the war would have taken quite a different course.

In the evening our cars fought their way through a thick blanket of snow to Breslau, where a short reception for members took place in the hotel followed by a eurythmy performance in the theatre.

During supper the conversation centred round the anxiety about the day's journey. Rudolf Steiner approved of the fact that his wife would not cancel the performance in Prague, and Carl consulted with our reliable chauffeur. My son Wolfgang and I were to accompany Frau Steiner on the journey across the mountains to the Czech border.

The next morning our first glance was directed out of the window. The snowstorm had abated, but the snowflakes still fell silently hour by hour, getting ever thicker. We set off first. The car slowly advanced through the deep snow. Marie Steiner, Mita Waller and I were stowed away and warmly covered up and Fräulein Pietsch handed us the food. Anxious disquiet and good wishes showed on every face as Carl closed the car door.

For the first hours the journey was quite comfortable. We saw where Bad Landeck lay and Marie Steiner told us about the time when she was there with her parents for treatment. Then the road became steeper, chatter ceased and the car began to struggle against the masses of snow and the bends. But by lunchtime we arrived safely at Habelschwert. High walls of snow rose up on either side of the street—how would we ever cross the pass? The inkeeper advised against it. Wolfgang and the chauffeur sceptically and nervously tested the chains again. Then we set off. The street was now frozen beneath a fresh fall of snow, so that we skidded dangerously in spite of the chains. A wild snowstorm swept across the summit of the pass. Metre by metre the car crept forward, till it suddenly sank into a deep snowdrift. We were stuck. Then a young man appeared, pulling his sledge behind him. We bought it from him. Frau Dr Steiner was put upon it. Wolf harnessed himself at the front and Mita Waller and I trudged along beside. We had to leave the car and the chauffeur to their fate, for we wanted to catch the only train to Prague at the Border.

At last we saw a logger's cabin through the driving snow and after hurried bargaining we got three little pony-sledges with their coachman, who would take us to the Bohemian railway station. Wolf remained behind to help the chauffeur. It had grown dark, the ponies were wild and restive. It was a strange journey.

Nevertheless, we got to the train on time—and then I stood alone in the little station at night. A single sledge was about to set off. I quickly asked if the man could take me with him. 'Yes, across the mountains to...' I did not understand what he said, but sprang aboard and the man whipped up his reluctant horse with a 'gee-up, gee-up' to get it to set off into the darkness. At last we stopped at a lonely house. The man told me to dismount, as he was going in another direction. I knocked on the snow-covered window—everyone was asleep. At last a surly woman opened the door. When she grasped the fact that I wanted to travel further she told me that her husband was out, the servant-boy was asleep, and she was unwilling to give me their only horse. Next I offered to pay many times the normal price in ready cash. Then she wakened the servant lad and when he had grumblingly harnessed the horse she called out: 'Get going before my old man sees you!' But just as we were going he arrived and screamed at us angrily: 'You must be mad to take the horse out of the stable at night. That's not on!' But when he heard what I had paid, he did not want to return the money, so he called out: 'Na, off with you then!'

In Habelschwert Wolfgang was standing in front of the Guest House, looking out for me. But, after a warm meal, we wanted to set off straight away in the car before the roads got still more snowed up. And this night journey was really horrible. The chains came loose several times. We often had to drive without a light because of lack of carbide (that was how it still was in those days) and then we ran out of petrol. We had to stop in 30 below zero in the dark (there were no heated cars then), and fill up the tank in the blizzard as the cold seeped into our bones. We arrived at Koberwitz at about 6 am, but still alive.

Rudolf Steiner also got to Stuttgart—nothing had to be cancelled. When I saw him there again in the spring he greeted me with: 'There you are—you nearly perished! My journey from Koberwitz was not very easy either—we had to keep stopping on the way to give the car hot poultices! I told my wife straight away that they would never let us out of Silesia again!'

Rudolf Steiner's second visit to Koberwitz 1922

The whole house rejoiced when Rudolf Steiner again accepted an invitation to stay with us. We met him at the station in Breslau on 5 May and he was greeted with enthusiasm by all as an honoured guest. He was on his own, for Marie Steiner was on tour with the eurythmists.

The conversation at table centred round the Breslau lecture and Rudolf Steiner recounted something of his lecture tour through the German cities.

He appeared at the stroke of 8.30 next morning. Breakfast had been laid in the bay window with its view over the park. The discussion turned to the ending of the Threefold Movement and I asked Rudolf Steiner if Carl's efforts and work on it had all been in vain. 'Oh no,' he answered, 'he will pick it up again in his next life on earth.'

I saw when Rudolf Steiner dipped his honey-roll into his coffee cup, that it would be easier if the cup had been placed directly in front of him, so, like a good housewife, I pushed the cup in front of him. But he interrupted his conversation and, with a triumphant smile in my direction, he moved it back again to its former preferred position. Carl laughed at me about that for a long time afterwards!

As Rudolf Steiner was leaving the room he paused by a table on which some books were lying. He turned over a few pages of *The Symbolical Figures of Christian Rosenkreuz* and remarked: 'One can learn a lot about the Trinity from that.' Next to it lay a book about the excavations of Tutankhamun's tomb. I asked if it were possible that the excavators dying one after another could really be the result of an ancient curse. 'Yes, of course,' said Rudolf Steiner, 'why should an Egyptian Pharaoh not utter a curse on those who disturbed his peace?' I said doubtingly: 'After so many millennia?' I was met by a glance from Rudolf Steiner—which at the very least conveyed to me that I should not ask such foolish questions—as though the spiritual force of a curse could be dissipated like dust in the wind!

Then he started to speak about the Görlitz Branch where he had been invited to the unveiling of the Jacob Boehme statue. He told us that Marie Steiner had upset the Mayor quite considerably at the time, when she told him in her impulsive manner that it was more a statue of Shakespeare than of Jacob Boehme. But, Rudolf Steiner said, she was right in one respect, as Boehme, Shakespeare, Bacon and Balde actually all looked similar—they had all had the same initiation teacher! I asked if this 'initiator' had been a bearer of the Holy Spirit. 'Yes, of course' he replied, 'Boehme himself describes how he had met with the Holy Spirit in his youth.'

On the journey by car to Breslau we looked at our fields right and left and I remarked that we were going to have a drought again. Rudolf Steiner said: 'We shall have to reckon with droughts everywhere. In this strip of land, however, it will be a little better.' And after a pause: 'Large estates in any case will find it difficult in future'.

On this day Rudolf Steiner gave his main lecture in the concert hall, and in the evening he gave one to members which, unfortunately, was not taken down in shorthand. I was deeply impressed by his words on that occasion about 'creating out of nothing', about which I was able to question him next day.

Then we stood at the door to say goodbye. Our Paula, who had been proud to look after Rudolf Steiner's rooms on her own, brushed past me with his cases. Then Rudolf Steiner appeared, ready for departure, and said: 'That maid with the skinny arms has carried my heavy suitcases all by herself. She must have nearly wrenched out her arms!' Then he offered his hand to each of us with the words: 'Next time I shall stay longer.'

The Agricultural Course 1924

The Whitsuntide Agricultural Course in Koberwitz actually came about through my husband (who administered the *Kommenden Tag* properties) having noted the specialist and unbelievably expert knowledge possessed by Rudolf Steiner. From this observation the spontaneous wish arose in him to beg Rudolf Steiner to enrich our knowledge of agriculture too, and to open new ways for us. I remember how, in his gay and loveable way—filled with the deepest reverence for Rudolf Steiner—Carl looked at me with his kindly eyes and remarked: 'I do not in the least see why I should not petition our Doctor about agriculture, if he gives such marvellous help to other specialists!'

Stegemann, Vögele and Erhard Bartsch had also begged Rudolf Steiner to give advice about agriculture. Carl had collected these endeavours together into a big request which he presented to Rudolf Steiner. At the next meeting Rudolf Steiner approached the agriculturalist Ernst Jakobi and my husband, and told them that he intended to hold the desired Agricultural Course at our estate in Koberwitz.

Correspondence on the subject then passed to and fro with Dornach, but no one could gather from it when Doctor Steiner would arrange his journey to Silesia. Christmas 1923 approached and still the time had not been fixed. Then Carl sent for Aki, his nephew and colleague, and told him he must find out the date for the Agricultural Course when he travelled to Dornach for the Christmas Conference. This he did in his usual efficient manner and returned bringing us the news that it would take place in Koberwitz at Whitsun. And so the happy preparations immediately began for this wonderful karmic event in our lives.

Then came the telegram to say that Rudolf Steiner was to arrive in Breslau the following evening and we all looked forward with joy to the coming event. Everything was ready to make these Whitsun days a festive occasion. Cleaning and renovating had been going on for weeks; there were still flowers to arrange and birch branches to place in the anteroom and in front of the house doors. In thankfulness, I played a chorale which rang through the house.

On the following evening many members had assembled at the railway station in the hope of receiving a greeting or a handshake. I was shocked to see how wretchedly unwell the Doctor looked—quite different to his usual self. Reports had reached us from Dornach about the serious ill-health of Rudolf Steiner, so that the greatest care seemed to be called for in every respect during his stay with us here in the East. These reports led us to believe that Rudolf Steiner's existence on the physical plane had been rendered almost impossible.

On our arrival home Carl summoned our young people and told them and Herr von Koschützky that a quite special watch must be organised, because it was known that attacks had been made on Steiner's life. The personal guarding of Rudolf Steiner even during the night was then assigned to Aki and a few of the young men. It now became clear to us that we were answerable here for his life.

It seemed to me that the Doctor had risen in his spirit into yet higher heavenly regions and thereby greatly increased the gulf between himself and mankind. It was shattering to see him—all had this same impression.

Rudolf Steiner told one or two jokes during the evening meal, to lighten the atmosphere. But things did not cheer up, even though the heartfelt wish of most of those present—to be close to Rudolf Steiner—had been fulfilled.

I had asked Frau Walter, who had formerly been in the Doctor's house in Berlin for years, to relieve me of the responsibility for his personal care, so that the meals could be prepared from the recipes of the Arlesheim Clinic. I could thus rest assured that Doctor Steiner would be looked after in this respect with the utmost care.

Graciously, but infinitely tired, the Doctor took his leave of us to go to his room.

Saturday, 7 June 1924
Reception of all the course participants, most of whom came from Breslau by the morning train. There were about 130 people, and several railway carriages were reserved every morning to bring them here. Sunshine lay over all, over the house, the park and the flower gardens. It was real Whitsuntide weather.

Herr and Frau Dr Steiner had breakfast alone in their rooms. They were staying in our bedroom and the workroom which lay quite shut off from the rest of the house. Dr Steiner came down the stairs several times to give instructions. Once I also saw him go upstairs to his young escort, Günther Wachsmuth. The latter told me later that he was still fast asleep when Rudolf Steiner entered.

It was a great pleasure to see all the happy faces and the welcome friends entering our house to partake with us in all these important events. Rudolf Steiner and Carl stood outside at the side entrance to receive the cars arriving with the distinguished guests, above all her Excellency Eliza von Moltke.

In the meantime the lecture-room had been got ready. In the dining room and vestibule 130 chairs had been set out, and the lecturer's desk stood in the centre between the two rooms. For the last six weeks the second wing of the mansion, which my brother had formerly occupied, had been put at our disposal, so that the great hall and the stairwell made an excellent waiting room for our guests. The entrance doors stood wide open on their arrival and until the beginning of the lecture people streamed straight out into the park and gardens as far as the log-house by the lake.

The conference began with my husband's words of greeting and his thanks to Rudolf Steiner for having come. The eurythmists were also mentioned.

Then Rudolf Steiner found some very warming words to say to us, which are largely recorded in the 'Agricultural Course' transcript.

The lecture which now followed is known from the shorthand report. It made one conscious of a burden of responsibility: that we were here entrusted with guidelines for the development of quite new research and knowledge, which the earth needs to renew its life.

At 11 am there was a mid-morning break. The whole audience repaired to the vestibule where a guests' buffet was provided for them. Great salvers laden with open sandwiches stood on the tables, pails with hot sausages and great cans of milk and fruit juice were provided and consumed with a hearty appetite.

After the break there was discussion at which questions could be asked. Lunch followed. Things had been organised in such a way that the lecture hall could quickly be transformed into a festively arranged dining room. Herr von Koschützki went round the house like a majordomo calling the guests to table. People were invited in turn to sit at Rudolf Steiner's table, for if more than twelve sat at one table it was not quiet enough for conversation to be held. Grace before meals was said by Dr Wachsmuth: 'The plant seeds are quickened in the night of the earth...' and Dr Steiner gave the Amen with a worthy dignity.

In the afternoon I drove with Marie Steiner to Breslau to inspect the theatre for the eurythmy performance. Later we were joined by Rudolf Steiner, Carl and my son Wolfgang. Rudolf Steiner examined everything thoroughly. Then, after a quick meal, we went to hear the karma lecture for members.

Late in the evening when the cars had returned to Koberwitz, supper had been laid for Dr and Frau Dr Steiner in their room, for them to eat alone. We others sat down together in the dining-room downstairs and each quickly got some bread and butter for himself without any ceremony— then Rudolf Steiner came downstairs and told us that he did not wish to eat upstairs on his own and that he was not a prisoner. Then he put us all in a happy mood with his jokes. Someone ran to fetch his meal and we sat together with him for a happy hour around the dining table. It seemed that he had become more cheerful.

Whit Sunday, 8 June 1924
This was Whit Sunday, with all the joy of spring. Sunshine within and sunshine without. Rudolf Steiner wanted to see the farm, the estate and the park on this day. Before breakfast I played *Now thank we all our God* on the harmonium, so that the whole household should begin the day with a feeling of devotion. Shortly afterwards Dr Steiner came out of his room. He fetched Frau Dr and we went on foot, first through my vegetable garden. Rudolf Steiner remarked: 'Everything here seems in such good order.' It had not escaped his notice that I had had all the paths strewn with fresh gravel. At the little arched gateway leading from the vegetable garden to the flower garden he observed with interest my attempts with the blackberry plants. We went along the garden paths which were bordered on both sides by broad beds with flowering bushes and roses. Dr Steiner thought that one could see from both the roses and lettuces, that they did not do so very well on account of the iron in the soil. Carl replied that everything here was suffused with iron, and that our water was barely drinkable. A purification plant with two filter cylinders had been installed, from which Rudolf Steiner tasted the water.

He then walked back to the house between Carl and me and said something to us which at first sounded strange, namely that the iron here in the ground had a connection with us. In yesterday's address he had spoken, not without reason, of the 'Iron Count' and the 'Iron Countess': 'You have both, in fact, an iron will—I mean in pursuit of your aims. The iron in the soil, and the iron will are connected, they have an attraction for each other.' It was only gradually that we sensed the meaning of these words.

Next we proceeded towards the farmyards. Frau Dr Steiner related to us how she and her brother had managed an estate in a wild and uncultivated part of Russia; they had done everything themselves except milking and carting manure.

Our cows seemed very pleased with our visit. In the meantime the fourteen young eurythmists, whom we had invited to a meal, had arrived from Breslau. The new-born piglets were their greatest delight and Rudolf Steiner also found them amusing.

Now came a walk through the park. In the bathing house by the great pond, which was entirely surrounded by dark fir trees, Rudolf Steiner sat down for a moment to absorb the lovely picture. At the Ranger's Pond we rejoiced at the sight of a duck with many ducklings. The happy laughter of the eurythmists sounded to us over the water—they were in the boat; Doctor Steiner loved them as though they were his own children.

A great festive table with many flowers was laid for thirty-six people for the midday meal. The young people sat in the vestibule, the folding doors of which were opened wide so that it was made into one big room. A new sunny cheerfulness radiated from the Doctor over us all. Then he stood up and made a speech about our house. We were quite embarrassed at being so highly praised. He spoke of the very congenial reception that everybody had received. After these words I think I heard nothing more of the speech. Caroline von Heydebrand supplied me with the following account from memory later:

'My dear friends, when the Keyserlingk's nephew came to Dornach we were worked upon with special magnetic force by what proceeded from him through his connection with Koberwitz. In the first place our eurythmy teacher was drawn to Breslau by this magnetic force. When we ourselves were drawn to Koberwitz we at first had many difficulties to overcome, but to our deep satisfaction we have been able to comply with this summons. Here one experiences the earth and the water as being entirely permeated with iron. And so now we can also fully understand the strong magnetic force which issues from here. It works however, not only in the soil and in the water, but also in the will forces of Count and Countess Keyserlingk. And through this will they have worked to make the arrangements here possible in so delightful a way.

'In our News Sheet I report about things that are happening [in our movement]. Now I shall have to report that a whole swarm of locusts has poured itself over Koberwitz. This swarm will certainly consume everything, and so I shall have to represent the Countess and the Count as being quite emaciated after the conference is over. But since the iron will is so strong in them, in future when we speak of them we will call them 'The Iron Count' and 'The Iron Countess'. I think that I speak here in the name of you all, when I cordially thank them for receiving us here with so much goodwill. I beg you to fill your glasses and drink to the well-being of Count and Countess Keyserlingk, that in the future too they may make their iron will active for the benefit of our anthroposophical work.'

At 5 pm the cars drove back to the town. The farmers had a meeting in Breslau at which Herr Stegemann wished to give an account of his experiences. Rudolf Meyer, the priest, told me that Carl had argued with Herr Stegemann.

Frau Steiner and I drove later to the town and picked up Rudolf Steiner and Carl to take them to the concert hall where the lecture was to be held at 8 o'clock. I questioned Carl very reproachfully in the car; I heard that he had treated his guest very badly. But in the middle of it Rudolf Steiner broke in with much amusement, remarking to me, 'Oh I thought the way he behaved was delightful.'

I heard later on that the two men had flown at each other in fullblooded farmer-language. Rector Bartsch had wished to mediate between them, but Carl had declined, saying that farmers got along all right alone, they understood each other's language.

Apart from his public addresses to the young people on 9 and 17 June and the two Class lessons, there was also an esoteric meeting on the morning of Whit Sunday after a stroll in the park and gardens.

Hardly anyone remembers about these. Rudolf Steiner spoke about meditations which the farmer should address to himself and the earth, and about beings which descend into the community of a farm and are active in the earth, the plants and the farm surroundings—and how then it would be possible for people to influence the weather by means of their moral will-forces. He spoke urgently about the degeneration of foodstuffs and how necessary it is to cultivate new plants. He also said that a whole new science should be established which would be effective not through itself but through esoteric realities. As I said, I was unable to obtain further details from anyone, so I shall just note here what I can remember from reading the lectures given in Berlin in November 1908 and from Dornach on 22 July 1924, and what Rudolf Steiner had also indicated to the few farmers on Whitsunday. He spoke about angel choirs which congregate above the places where people form communities, to prepare for the in-streaming of exalted spiritual beings willing to help mankind. By forming such communities one would be able to gain access to those hierarchies which are served by the nature spirits, which would in their turn be able to exert a healing influence on the life of plants. And Rudolf Steiner added, with a friendly glance towards Frau Dr Vreede: 'And it is Dr Vreede who will bring about this connection to the spirits of the spheres through her work.'

In November 1908 Rudolf Steiner said the following in his lecture in Berlin: At present people only agree about mathematical truths. Whoever has once accepted them knows that it has to be thus: 3×3 must equal 9. The only place where disagreement can occur is when truth is dimmed by passion, by sympathy and antipathy. A time is coming when mankind will more and more be taken hold of by a comprehension of the inner world of truth. When that time comes agreement will prevail over all individuality. Ever greater peace will be achieved when truth, as such, is understood.

When people voluntarily allow their feelings to flow together, then something will be formed which goes beyond humanity. When people share activities together in freedom, then they congregate about a centre. Human feelings which converge on a central point enable exalted beings to work like a group-soul. These exalted beings form a new kind of group-soul which is compatible with complete freedom of the individual, but their activity is dependent upon human unanimity. It will lie in the capacity of human souls to determine how many of these higher beings will be able to descend. The greater the disparity in mankind, the fewer will be the number of higher beings who can descend. The more we can band together voluntarily into groups, the more will exalted beings be able to work among us and the quicker will planet earth become spiritualised.

In later times people will live in connection with one another in conditions which they themselves will have brought about. They will band together into groups according to their different points of view. There they will be able to develop their individuality in complete freedom, and there will be many such associations in future.

On 22 June 1924 in Dornach, Rudolf Steiner said similar things to those he addressed in an esoteric gathering after the Koberwitz Course, which the following approximate words convey: 'I now have to make an introductory remark, which has to be taken seriously. What has been said in talks since the Christmas Gathering can only be divulged to audiences by reading the exact text. It cannot be conveyed in a free rendering. I would have to oppose that!

'In the case of such weighty matters every word and every sentence has to be carefully considered. A genuine common spirit must enter the whole anthroposophical movement, otherwise we shall fall into the mistake into which a number of our members have fallen.

'These members thought that anthroposophical wisdom had to be scientifically adapted. And we have seen how much harm to the anthroposophical movement has thus been "achieved"—I put it in inverted commas. What I call responsibility towards the spiritual world begins to take effect to an outstanding degree. I believe that these words I have just spoken will be understood.'

This serious warning refers to the fact that during the Whitsuntide gathering in Koberwitz Rudolf Steiner pointed out that only the Word, as it is spoken of in St. John's

Gospel—consistent with the Logos, the Universal Word—is fit to transmit the truths and knowledge of anthroposophy. Every adaptation in the sense of present day materialistic science would sunder the esoteric messages and their effect from the co-operative activity of the Logos.

The following days of the course brought a repetition of the same order of events. Happy people streamed over early for the lecture. After a very stimulating mid-morning break, discussion followed. Then came the midday meal, and, without an interval, the drive to the functions in Breslau from which we only returned at 11pm each night.

The whole house sparkled with cheerfulness and the happy activity of all. In such joy one feels that there can be no more sorrow either now or in the future—just as in sorrow one feels that there can never be any more joy. Rudolf Steiner himself was completely changed; he seemed to give the lectures with great pleasure and looked quite youthful. From day to day he looked better in health and more cheerful, and his health and good spirits spread confidence and happiness over all who were present. After his first refusal to eat alone in the evening it was decided that he should always have supper with us. The whole household looked forward to this and anticipated the evening meal with pleasure. Doctor Steiner too was most charming, and laughed with everyone with such sunny cheerfulness that all were affected by it and went off to bed in utmost happiness.

At the midday meal we came to speak about the newly published autobiography of Rudolf Steiner. He turned to me in his very loveable way and asked what had pleased me most in it. I answered that it was the story of the three boys whom he educated in Vienna. They were to be envied. He said that there had really been five, occasionally still more. 'I was on a special footing with one little boy who was five years old. He had so much energy in him that he did not always know what to do with it and was forever complaining. Once when I went to him he thought hard for a while, but no expression seemed to satisfy him; at length he said: "You are as stupid as three donkeys!" But he was a splendid boy. And there was an Uncle Eberhard', he went on, 'with whom I was on excellent terms; he was much embarrassed once when the little fellow called out to me in front of him, "Uncle Eberhard said that you are the most slovenly person he has ever seen". "Yes, and why so?" I asked him. "Well, your tie is always crooked!".'

Frau von Moltke told me later on that Rudolf Steiner had often spoken about these times to her and how, as a young man of twenty one, he had been the teacher of a handicapped pupil. Steiner had been able to observe serpentine formations which issued from the child's astral body. He had only consented to undertake his education on condition that he was given sole charge of the child. So for six years of his life Rudolf Steiner had been a curative educator. During these years he so changed the boy that he had been able to take his matriculation and become a doctor.

Because of his ailing digestion Rudolf Steiner had everything specially cooked for him. On one occasion, however, the old butler offered him a dish with some appetising ham, thinking that anything so good could not harm him. Doctor Steiner declined laughingly: 'No, no one can demand this of me!' At almost every midday meal he had an excellent soup of vegetables, herbs, cheese and caraway, then two fresh, lightly boiled eggs and an Austrian sweet made from flour. He always ate up all that was given him with seeming relish. Once I noticed with surprise that he let the soup stand. I asked him why he did not eat today. 'Because I have no spoon' was the laconic answer.

Once, when there were strawberries for dessert, I asked him if it were really right to cultivate strawberries to such a size and he replied that under certain circumstances strawberries could die out altogether through such a practice, because the aura of the group-soul was disturbed by it. I was reminded of our old avenues of poplars which Napolean had planted to mark out the way to Austria for his army and I asked him whether the dying out of poplars was connected with the same thing. 'Yes,' he replied, 'the mother tree in America from which these were grown has died.'

There would be much one could still relate from these interesting meal-time conversations, which Doctor Steiner also listened to with appreciative interest.

Carl once said: 'One thinks one knows something as a farmer, but one stands as stupid as a school-boy before the vast expert knowledge of the Doctor!' Then he thanked Rudolf Steiner for what he had given us hitherto and he begged him to help agriculture still further. I was somewhat horrified at his having the courage to tell the Doctor that he must consider himself as our Head-farmer—but the latter replied that he was pleased by this, for he was descended from farming-stock.

The farmers had actually formed themselves into an experimental circle, under the supervision of Carl and Herr Stegemann. Rudolf Steiner spoke the following words to them, which cannot be taken seriously enough since they were entrusted to this young branch of anthroposophical science:

'First of all let me express my deepest satisfaction that this experimental circle inaugurated by Count Keyserlingk has come into being to widen agricultural interests. To begin with Herr Stegemann related some of the admirable experimental efforts he has been making on his farm. From this arose a discussion between our excellent Count Keyserlingk and Herr Stegemann which has led to the framing of the resolution that we have heard today.

'But we must avoid the kinds of mistakes which have only become obvious in the course of time. It is naïve to imagine that a professor or other scientist of today can suddenly be won over to anthroposophy. That does not happen. People would no doubt like to have anthroposophy, but they cannot endure—and rightly, as I have admitted—a cobbled together, muddled mixture of anthroposophy and science. There is no future on those lines. And therefore I welcome with great pleasure the initiative of Count Keyserlingk, through which it has been decided that the community of agriculturalists should unite with what has been founded in Dornach as the Natural Science Section. That resulted from the Christmas Conference, as does this which now comes about. Thus whatever comes from Dornach will be what is needed.

'However I naturally cannot agree with Count Keyserlingk's suggestion that this group of professionals shall be merely an executive body. Indeed the foundation for the work that we shall have to do in Dornach must, from the very beginning, come from you. We shall therefore need active, most active, fellow-workers from the outset, not simply an executive organisation. Now, if you will allow me one suggestion—which was discussed between Count Keyserlingk and myself several times during these last days: an estate is always an individuality in the sense that it is never really the same as another estate. Climate and soil conditions afford the most radical basis for the individuality of an estate. An estate in Silesia is not the same as one in Thüringia or in South Germany. They are actually individualities.

'A great deal has been said—but in all goodwill and without irony, since people have rather enjoyed it—about the differences of opinion at the first meeting between Count Keyserlingk and Herr Stegemann. I almost thought from what arose then that one must consider whether the anthroposophical *Vorstand*, or someone else who could be present every evening, should keep the combatants in check. But I have gradually become convinced of something quite different. What came to expression in this difference of opinion is really the foundation for an inner tolerance among agriculturalists—an inner forbearance among colleagues. It is only the outer aspect that appears rough, but underneath is a certain tolerance; so that I may once more express my deep satisfaction with what has come to pass here through you.'

It was remarkable that during all these days Rudolf Steiner fell into line with all the timing and practical arrangements. This was a lesson to us about how, in working together with others, one must not disturb the plans and arrangements these others have made, but must learn to subordinate oneself. Carl was responsible for the daily timetable, and was very anxious that Dr Steiner should not get overtired, so he asked him if for once he would take a thorough rest. 'Oh no,' was the reply, 'that would only tire me.'

Rudolf Steiner slept very little. The bedroom was so arranged that a large table was drawn up to the bedside, on which books and papers could be placed. Guests told me later that they had often looked up at Rudolf Steiner's window at night and that there was always light burning. And when one saw the pile of post that was collected from him every morning around 5 am by Herr von Grunelius, one can imagine that he used the silence of the night not for sleep, but for work. The articles for *Das Goetheanum* had to be sent off early, and Frau Dr Wegman told me that Rudolf Steiner wrote to her every other day. In the night of 14–15 June Dr Steiner, at the request of the priests at the conference, wrote the new text for the Consecration of Man for St John's Day. Rudolf Steiner set us a daily example of how we should live in the present and have absolute trust in the verity of the spirit in earthly life.

And now I have to report something which caused Rudolf Steiner a lot of pain during these days. One evening, after the esoteric class in Breslau had already begun, and the doors had been shut by the doorkeepers, members who were late knocked loudly and repeatedly on the door. The doorkeepers opened it. Herr Doctor had to interrupt his words. At the end of the class something happened which will certainly be remembered to the end of their lives by all those concerned. Carl had to appear on the platform and announce that the doorkeepers and the members in question were to report to Doctor Steiner with their cards. He took their class membership cards from them. Through that the seriousness attached to membership of the class—right into the profoundest depths of our karma—was demonstrated to all of us. Dr Steiner gave Carl instructions that the younger members should be re-admitted fairly soon, the older ones later.

Rudolf Steiner played an active role in all that happened in Koberwitz and the Breslau Branch and it is difficult to pick out from all the many records of those days what will later appeal to those who only know him from his works. Therefore I will mention those little things which so endear us to the engaging figure of Rudolf Steiner.

We had invited several gentlemen to lunch who had often made it difficult for Carl as chairman of the Breslau Branch. Afterwards in the car Rudolf Steiner leaned over to Carl and said: 'Were those by any chance Your Majesty's most humble Servants of the Opposition?'

One morning Rudolf Steiner stood in the hall waiting for me. He wanted to apologise to me because his fountain pen had leaked and the ink had gone onto the pillow. I could only laugh and say: 'But, Herr Doctor, that is charming!' Later Rudolf Steiner announced: 'The Countess is no Philistine; I made ink-spots on her beautiful linen and she was very pleased!' Then Wachsmuth said: 'Let's hope they aren't like the inkspots at Wartburg castle!'

There was some hitch in the eurythmy performance planned at Görlitz and it was cancelled. After this decision Dr Steiner turned to me and said in a serious tone of voice: 'To begin a thing without finishing it is worse than if one had never begun it.'

Another saying of Rudolf Steiner which occupied me deeply later on was the following. Herr Winkler had asked Dr Steiner if he would give us some information about the previous incarnation of Caspar Hauser. The following day Rudolf Steiner brought him the answer: he had done spiritual research both where Caspar Hauser had entered physical existence and also where he had been murdered, but he could not find any indication of a former or a subsequent incarnation. He was a higher being who had a quite special task on earth.

On one of the following days we had a look at the Kreuzkirche and the Sandkirche in Breslau. Frau Dr Steiner and Dr Wachsmuth sat together in the front row of seats. Rudolf Steiner stood for a long time absorbed before the altar. I was further back alone by a pillar. Suddenly I had a feeling of being completely forsaken.

Dr Steiner turned round and came to me through all the long rows of seats. He sat down beside me for a moment and then made his way back through the length of the nave to the high altar. That was Rudolf Steiner's goodness of heart.

On the evening of the first lecture for members of the Breslau group, Rector Bartsch gave Dr Steiner a kindly welcome and said that his coming had been awaited like that of a father. Doctor Steiner thanked him in a friendly way, but declined to belong to the older generation. Then, suddenly and unexpectedly, he stood up and threw into the meeting the words: 'I have decided to become young again.' These words had a note of enigmatic and shattering profundity, for in his features one could read his approaching death. They had the effect of being parting words to the friends, as if to say: 'I go towards death, in order to become young again, to be born again.'

This first lecture was an introduction to the subject of the following lectures: karma, and how destiny leads through death and birth and comes to expression in the earthly realms between birth and death. Rudolf Steiner spoke of how, in numberless cases, one can see that difficult, sorrowful, tragic experiences come to good people, whereas those whose intentions are by no means good meet not with bad, but agreeable experiences in life. While we see how necessity is revealed in natural occurrences, as effect follows cause, we cannot see what is woven into the spiritual connections of our moral life. For how many people has this question led to despair and rejection of the divine—and for how many could these karma lectures give new courage to face life!

For Sunday afternoon a social evening was planned at the 'Mathiaskunst' at which the Silesian members might meet Dr Steiner informally and get to know him personally. The 'Mathiaskunst' building lies among old trees on an island in the Oder. It was a cheerful, jovial excursion when the young people, the kitchen personnel and necessary staff, with the whole store of provisions piled high in cars, all drove off from Koberwitz to Breslau. After several speeches a long desired wish of many hearts was fulfilled as Rudolf Steiner made himself wholly available to anyone who wished to speak with him. Many had come to ask him for advice, among them a blind lady, and I can still see Rudolf Steiner's gesture as he took her groping hands lovingly in his clasp.

Still another picture: an excited young man is talking to Rudolf Steiner in an excited and apparently endless way. Herr von Koschütski appears behind Rudolf Steiner's back, holding up his watch as a hint to the young man to bring his talk to an end.

Rudolf Steiner was certainly pleased at the love and unlimited trust which was shown him, yet at the end he appeared exhausted and we were glad when this successful evening could be brought to a harmonious end with words of thanks and flowers.

On 16 June the eighth lecture took place, the last of those held before the assembled farmers. Carl had come to an agreement with Dornach to keep Dr Steiner's indications secret until research work had produced results which could be made public. Those involved were committed to keep silence until then. Rudolf Steiner therefore concluded his lecture with the following words: 'I am in entire agreement with the strict resolve which has been made by our farmer friends here present, namely, that what has been given here to all those sharing in the course shall remain for the present within the circle of farmers. They will enhance it and develop it by actual experiments and tests and then the farmers' society—"The Experimental Circle" that has been formed—will fix the point in time when, in its judgement, the tests and experiments are far enough advanced to allow these things to be made public. Full recognition is due to the tolerance which has allowed a number of interested persons—not actually farmers—to share in this course. They must now recall the well-known opera, and set a padlock on their mouths; and not fall into the prevalent anthroposophical mistake of straightway proclaiming what they have heard far and wide. We have often been harmed in this way. It makes a great difference, for example, whether a *farmer* speaks of these things or someone who is remote from farming life.'

Regrettably this warning about not divulging as yet

untested statements was not followed. Many, also young farmers, blatantly and enthusiastically spread the news about—which was the source of serious internal differences of opinion, and later of great difficulties.

Then Rudolf Steiner expressed his thanks in exceptionally kind words to our whole household and each single member of it who had been engaged in the preparation and running of this course.

I have taken down in my notebook some details which appear to be worth mentioning from the following days:

Dr Steiner had held a Class lesson [of the School of Spiritual Science] and had been obliged to refuse admission to one of the members, since his nerves were in a poor state. I said mournfully: 'He will now perhaps be weeping at home.' I received as answer a look from the Doctor which meant: 'Do not talk so foolishly! Everyone must bear his own karma!'

On another occasion when nasal operations were the subject of conversation, Doctor Steiner made the highly interesting remark that the action of the heart might be greatly affected if after such an operation too much or too little air were inhaled as a result.

Following on the conversation about the cow-horn manuring, I asked whether the stirring of this manure for our extensive acreage would not prove very difficult. 'Oh no,' he replied laughing, 'it is very easy in your case. You always have so many visitors here, so when they have drained their coffee cups, just put a bowl for stirring before them.' Then I asked how many cow-horns we should use and Rudolf Steiner said he thought that once things were established, it would not be so very many.

During a private conversation with Carl it was said that people did not take sufficient note of what was happening in the East. It was there that Europe's fate would be decided, for Germany would someday be only an American colony. I was not present during the much discussed conversation about Germany's future. My son, however, who was also not there, got the following report of it soon afterwards from his father and Frau von Moltke:

Carl had asked Rudolf Steiner what he foresaw would be Germany's future, and the latter replied: 'The [factory] chimneys will topple and Germany will be reduced to an agrarian state.' Then Rudolf Steiner said very seriously: 'All will depend on forming islands of monastic seclusion in the countryside where German cultural and spiritual life can be cultivated. Foreign lands will send their sons and daughters there to be educated.' After a pause he added: 'And it will be a long way from one island to the next.'

Rudolf Meyer noted this conversation in roughly the following way: Germany has ceased to be of political importance. It will be reduced to an agrarian state in which oases of spiritual life will be able to exist. Then anthroposophy will have to proliferate everywhere throughout Central Europe. And it will be capable of that. Germany could acquire a mission like Greece after its suppression by Rome—as the spiritual teacher of the dominant race. It only has to recognise its task, otherwise Europe will sink into utter barbarity, and culture will die.

In connection with this question about the fate of Europe, which is of such vital interest to all our friends, I should like to mention what Frau von Moltke said to me regarding the messages from her deceased husband, the Chief of General Staff, which were communicated to her through Rudolf Steiner in the following way:

From 1900 to 1914 souls were shrouded in a cloud of Luciferic feeling; and since that time until now it is an Ahrimanic one. Instead of there being any kind of order, Germany is now a refuse heap, which weighs like a stone on our hearts.

The spirituality of Europe in the 9th century was entrusted to Pope Nicholas and his counsellor. At that time the necessary spiritual streams came to Rome from the Odilienberg, for St Odilie was a light in the darkness of Europe, the effect of which has now been exhausted. A renewed connection to the time of Nicholas will only be possible at the end of the second millennium. Today hearts are empty and heads lack counsel. Whole cities will fall into ruin and only agriculture will remain. All will depend upon those people who have the courage to found monastic colonies in which spiritual life can be cultivated. Spiritual light will only make its appearance during this century if Michael succeeds in finding the way to the altar in the Astral Light, on which burns the flame which is dreaded by Ahriman.

I should like to identify the name Moltke in full consciousness at this point, for we have been warned by Rudolf Steiner that human beings who are the bearers of high aims can be reduced to impotence if their names are not handed down to posterity in a living way. We therefore bear the responsibility to see that the name Moltke remains rightly in people's consciousness, for it was certainly not by accident that these conversations took place in our house.

Frau von Moltke was conscious of the significance of her responsibility when she submitted what follows to print, so that responsibility for its content might be shared by its readers.

Rudolf Steiner gave a letter to Carl on the morning of his departure with a request to hand it to Frau von Moltke. This letter has been published by W.J.Stein in his book: *The Ninth Century, World History in the Light of the Holy Grail.*

Frau von Moltke added the following for Dr Stein:

'Dear Dr Stein,
I will gladly oblige in making the following note available to you for your book.

A conversation between Pope Nicholas and his counsellor, the Cardinal, which Rudolf Steiner conveyed to Eliza von Moltke on 17 June 1924:

Pope: Must we lose what spirituality has brought us, now that tidings of the Crucified One have brought heaven down to earth?
Counsellor: What is outmoded must fade; death is merely renewed life. I see the life of Europe rising out of Asia's decline.

Pope: It is a difficult decision.

Counsellor: Nevertheless it is required by higher powers, so that Ahriman is given the right direction within the soul-life which shall shine forth from Franconia to the East. 'Twas told me by the Northern Lights, which also possesses a soul, as I lay in my own country one bright summer evening and listened to the voice of Gabriel, who wishes to bring a New Europe to birth.

Pope: Are you sure?

Counsellor: There can only be certainty where higher powers are speaking, and I am sure that their message is clear.

Pope: Maybe they speak clearly enough, but I also know that the centuries to come will weigh heavily upon our souls.'

Questions of worldwide significance, encompassing broad spans of time, moved the hearts of those who were present at the Agricultural Conference; and they were answered by Rudolf Steiner.

Esoteric conversations

The Agricultural Course had come to an end. Now the young people came to Rudolf Steiner with their pleas. They trusted Rudolf Steiner—and one can say that Rudolf Steiner trusted them too.

The 'Free Anthroposophical Association' had been founded—free in the sense that the youth impulse should be left free to develop, and free from restrictions, so that the mighty impulses coming from the Michael Age could be taken up with fiery enthusiasm. For the the divine flame of enthusiasm for the advancement of humanity and of the earth lights up in those who seek the Christ within themselves. Therefore Rudolf Steiner ended his address to young people with the words: 'The sunrise must be a flaming one.'

This gathering took place on the day of departure and was therefore fixed for 7 am. Rudolf Steiner stood in the hall on the dot of seven and the young people had arrived too, yet the organisers were missing. My husband and I were very indignant—how could one keep Rudolf Steiner waiting! Eventually Carl went to the telephone and made a call to Breslau. It then transpired that those we were waiting for had overslept.

Carl and I learned a lesson for life from this: that, like Rudolf Steiner, one could remain good-tempered and friendly and not at all impatient or angry. He said: 'Oh, if one calls to mind that Grunelius collected my post at 5 am every morning, it is understandable that he occasionally oversleeps!'

At length they arrived and we collected in the hall for the last time, where the chairs were arranged in a half circle. The lecture was delivered in a joyful spirit, and yet there was something of deep seriousness present in the meeting, which we took to be the legacy given to our house and all who had been present in it, by the departing Rudolf Steiner.

The challenge had been given us during the Agricultural Course to place a spiritual science alongside the science of the universities, as a contribution towards overcoming the ever more threatening rise of materialism.

Carl's life soon afterwards ended in deep suffering, when working in this way was made impossible for him.

That must have been the reason for his unexpectedly quick withdrawal from the earth. The course was to have offered him the tools with which to sanctify the earth itself—may his successors inherit this feeling of responsibility with the same unquestioning faith! May the words of this address by Rudolf Steiner become active in your hearts:

There is something more to it than the mere forging of Michael's Sword. It is a fact that in the occult regions of the earth what is prepared by the forging of Michael's Sword is carried to a subterranean Altar in the process—to an Altar which is invisible and which really exists beneath the earth.

To become acquainted with nature-forces under the earth, to get to know the divine beings working in nature leads to an understanding of the fact that the Michael Sword, in the process of being forged, is really carried to an Altar under the earth. The dead take part in this. It has to be found by sensitive souls. It is necessary that you get involved and work together so that more and more souls find the Michael Sword. Nothing is accomplished by the mere forging of the Sword. It only becomes effective when it is found. Be strong and modest in your self-assurance that, as young people, you are called by destiny to seek for Michael's Sword, to find it and carry it out into the world.

At the close of the lecture Rudolf Steiner walked round the circle to say goodbye. He went to each one and not only gave them his hand, but took every hand that was offered him in both of his, as though sealing a bond with them. It was like a solemn commitment to a soul-pledge of faithfulness to serve the spiritual tasks of our time.

We did not suspect at the time that this handshake was for most of us a leave-taking from him for the rest of our lives.

These words were spoken to the young people because it is they who have forces for the future, impulses which penetrate into the soul and at the same time into the interior of the earth. For one's own inner being is also the interior of the earth, into which we can all penetrate. In this way the Altar will be found by following the path and teaching of St. John. Rudolf Steiner made this appeal many times and one can truthfully say that the Anthroposophical Society was on the point of failure through neglect of this advice. It is not head-forces but heart-forces fructified from the depths which will be able to lead us out of the present crisis.

According to Rudolf Steiner it is therefore not only a matter of developing the forces of iron within us, but of penetrating down into the depths of the soul, to the ground of the world where the Altar stands in objective reality. The forging of the Sword and the placing of it on the subterranean Altar was the deed of Michael working with the dead—but the finding of it must be done by the living.

It requires courage and truthfulness to pierce the depths of one's own soul-world. There one encounters the Apocalyptic Beasts. The abyss of one's own soul-darkness is revealed to one—and human beings are afraid of that. 'It is fear' Rudolf Steiner once proclaimed, 'which prevents man from penetrating into his own soul depths!'

If this warning had been heeded a true Rosicrucianism would have been able to oppose the storm of materialism, because, being deeply anchored in the fundament of the soul, it would have been linked together with the Godhead. People prefer, however, to persuade themselves that a command of material life is the most important thing. One therefore drowns the call of one's own soul, which sounds a warning from the depths—and one stays on the surface.

Frau von Moltke once said to Rudolf Steiner: 'People are unable to penetrate into the depths!' 'No,' said Rudolf Steiner decisively, 'they do not want to.'

Man's will has become free—he is able to decide for himself which direction to take.

The agricultural lectures were thus given a solemn ending. Rudolf Steiner had the kindness to come up to my room, where he spoke to me about the kingdom in the interior of the earth.

We know that at the moment when Christ's blood flowed down onto the earth at Golgotha a new sun-globe was born in the earth's interior.

My search had always been directed to the study of the earth's depths, for I had seen a golden kernel light up within the earth—named by Ptolemy the primeval sun. These golden depths I could only connect with that land which Rudolf Steiner said had been hidden from the sight of man, and that Christ would open the gates in order to lead those who seek for it to the submerged fairy-tale land of Shambhalla, of which the Indians dream.

What I had experienced had been so real—and yet I did not want to lapse into fantasies. I had written down my questions. Rudolf Steiner read them out aloud:

'Where can we find the land Shambhalla? I can imagine it somewhat as follows: in the beginning there was fire and light. These took their strength from one another—reciprocally in complete harmony. Then the light rejected the forces of fire; and the smoke which was thereby hindered from rising into the light, fought back and accumulated above the fire. Thus the primeval fire was gradually covered over by the ashes from the smoke. The ashes from the primeval fire form the mineral part of our earth.'

Rudolf Steiner reflected for a moment: 'That is interesting!' Then he continued to read aloud: 'First of all people could see through to their original homeland, the primal sun, but the earth became ever more dense—smoke increasingly darkened their view of the primal sun. Man appeared to be separated by a mineral layer from the primal sun, which is paradise, the land which was lost from the sight of earthly eyes, but from which man originated. The kernel of the earth consists of golden fire—around which a dark girdle is cast—the smoke of the mineral realm.'

Rudolf Steiner paused once more: 'What you have written

is correct,' then he continued to read: 'We human beings, however, have to rediscover the connection to our origin, otherwise we shall become more and more excluded from life and will solidify with the minerals of the earth. If my assumption is correct we must start to penetrate the depths to the golden ground of the world. The way leads through the darkness, through the layers of smoke which envelop the living fire. The earth-layers of the mineral world are permeated with the kamaloca of our soul-life.'

'Yes, that is correct,' said Rudolf Steiner and continued to read:

'There we meet with everything which connects the mineral darkness to the hell of souls. When we step out of the human world—which can occur even while the body continues to exist on earth—then we arrive in the land of the dead. When the dead person has accomplished his kamaloca, he rises to the starry worlds. I would think though, that if it were possible to achieve a state of sinlessness in kamaloca and, instead of rising upwards, one were to remain in the depths, would not the golden gates of the realm of fairyland open once more for the petitioning soul?'

Rudolf Steiner: 'Yes, that is possible.'

I: 'And is the interior of the earth made out of that gold which comes from the hollow cavity in the sun and which is destined to return there?'

Rudolf Steiner: 'Yes, the interior of the earth is of gold.' Then he put down the paper. Still I continued to question him for my assurance: 'Herr Doctor, when I am standing here on earth', and I pointed to my green carpet, 'then the golden land is beneath me, deep in the interior of the earth—and if I now attain to sinlessness and remain in the depths, the demons will not be able to harm me and I will be able to penetrate beyond them—and reach the golden land?'

Rudolf Steiner: 'If one passes through them accompanied by Christ, the demons will not be able to harm you—but otherwise they would indeed be able to destroy you!'

He added emphatically: 'They can, however, become our

helpers. Yes, that is so—the path is a true one, but it is very difficult!'

I knew now that my investigation had been on the right lines, but this path was a 'very difficult' one.

It was Hölderlin who had shown me his path. For years his astral forces had led me. I have climbed after him through the depths of fire and death until he found the way out to the light, the way out of imprisonment. He, who even left the earth in order to search for the Gods—'... for to be alone and without the Gods was death to him'—knows the Christ in the depths. So he does not rise up to the stars as other dead people do. No, he prefers to penetrate to the depths to the Sun Being at the centre of the earth.

The indication by Rudolf Steiner that what I had seen was the truth, but 'very difficult' left me wondering, and it seemed to me that he meant that there was a different way for the pupils of Michael.

Now it has been granted, through the help of Michael, that the Christ who appears in the etheric realm of the earth will illumine our eyes to see the lost magical land. Then this land will rise up to us and man will bind himself to a new worldday in the light of the sun, when he sets free from their enchantment the spirits of the elements who have darkened his vision and rendered the earth solid and opaque.

In Rudolf Steiner's Mystery Drama *The Guardian of the Threshold* Ahriman says: 'The gods, however, willed to rule on earth, and from their kingdom they did one day thrust my power into the depths of the abyss ... and thus 'tis only from this place I dare send out my powerful strength upon the earth. But in this way my power turns into fear.' And Capesius calls out: 'Oh, in the depths dark fear is threatening!'

The spirit-pupil must therefore be aware of the fact that whoever penetrates into the depths has to pass through Ahriman's kingdom, and the beings of that kingdom touch him!

As well as thinking and feeling the Being of Christ who united with the earth through His death, one must also strive to become a brother of the divine Christ in order to attain in a full and living way the rays of victory which await us beyond the darkness.

The spirit-pupil thereby gains a weapon to protect him: the reality and power of perceptive consciousness!

Perception of the laws of light is the shield which, in the hour of danger, is ready at hand. The spirit-pupil who would dare to journey on the road into the depths, must be in possession of that knowledge which ensures victory through Christ.

I then asked Rudolf Steiner if we would always be exposed to destruction by the demons: if they always have to destroy everything or whether we can be taken into the protection of Christ. He replied: 'When we have entered into the circle of the sun, the demons will no longer be able to harm us.'

My next question was also related to the primeval depths. I asked him what substance kept the primeval fire alight, for where there is fire there must first be substance, and so the primeval fire is not the beginning which produces original life out of itself.

Rudolf Steiner replied: 'No, it is different—it is not so in this case. On earth there has to be material for fire to burn. But the primeval fire burns spontaneously, it is its own substance and being! Substance only came later and was added to the Fire.'

I: 'Then is the primeval sun of Ptolemy, which he perceived in the centre of the earth as the creative ground of the world, the golden fairyland Shambhalla?'

Rudolf Steiner: 'Yes—and midnight conceals it.'

Then I presented Rudolf Steiner with the diagram [on page 88] and asked: 'Can one draw it like this?' and he answered: 'Yes, one can do that.'

'The festivals,' I said, 'are always at the points where a new spiritual current enters in and relieves an old one. The Father is succeeded by the Son and the latter is succeeded by the powers of the Holy Spirit.'

Rudolf Steiner then asked me what I called the three uppermost points and I answered that they were to be assigned to Easter, Whitsuntide and St Johns. Then I stopped

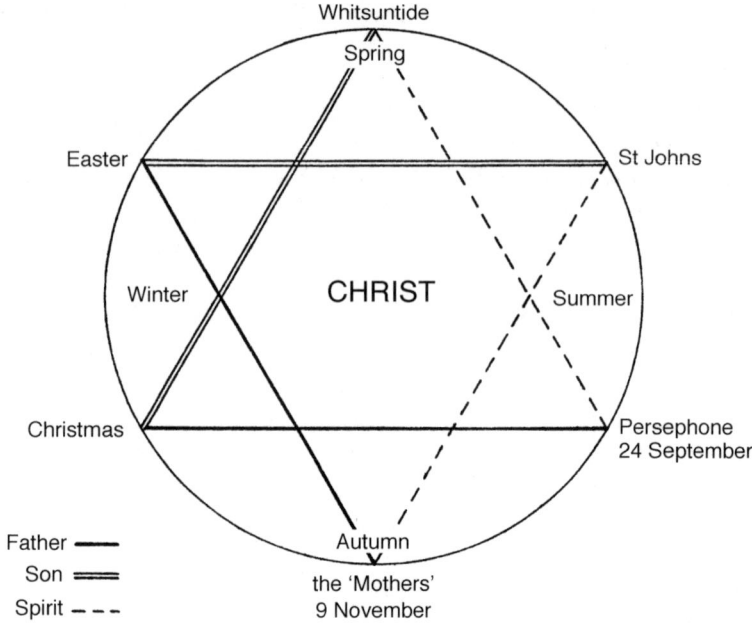

because I did not know what to call the lower point on the right and the one at the bottom. He pointed to the lowest apex: 'What do you call this?' 'The journey of Faust to the Mothers' I answered. 'Yes,' said Rudolf Steiner, 'on 9 November,' and added: 'And here, on 24 September, the Festival of Persephone was celebrated by the Greeks.'

I then put several questions which connected to what had been spoken about during the last few days, and which could not be separated from it in my view. I only wish to recount a few of these questions here, ones which might provide confirmation to one or other of us who are investigating these things further.

Rudolf Steiner asked me if I had any further questions and I showed him my calculations in astronomy and voiced my opinion as follows: 'A completely new astronomy is developing, which has nothing at all to do with what one calls astronomy today. I do not dare to work further on that until I know if what I am writing is true.'

Rudolf Steiner: 'If you experience these things as you have, then they will be true. You only have to be quite sure in yourself, quite selflessly true.' And, only audible to spirit ears, he added: 'Because these truths can only be experienced in selflessness.'

I: 'You told me Christ is the divine ego. So is the earth the physical body of Christ, the etheric world His etheric body and the astral world His astral body?'

Rudolf Steiner: 'Yes, of course.'

I: 'I see many things but, without having asked you, I surely cannot say, for instance, that in the slag of cooled moon-craters the dead fire-forces of God's adversaries are crouching and wrangling with Him?'

Rudolf Steiner: 'Why do you not like to say so, if you experience it in that way?'

When Rudolf Steiner was with us the previous time he had said something which has occupied me a great deal since: here on earth one can close an inner door, but in the spiritual world it is different. On Sirius one hears everything that people think.

I studied all I could find to gain a grasp of the identity of Sirius. Mythology only revealed that it was the star of Zarathustra, and stands as Isis in the constellation of the Dog. Neither did the Astronomical Course [by Rudolf Steiner] give me any information about the secrets of Sirius. Then I thought that I should tackle this question from a different angle and formulated it according to my own ideas as follows: Where is the centre in which all human suffering and all human joy in heaven and in the cosmos can be perceived? Then the answer sounded in my soul: it is His heart—the heart of Jesus-Zarathustra in the earth-depths of midnight. I asked Rudolf Steiner about this and he said: 'Sirius is the heart of Jesus-Zarathustra and is in the depths of the earth.'

'The heart is the deep' said Jakob Boehme: in the depths of his heart man also attains that other Heart, which would bear with him all his suffering and all his joy.

Then Rudolf Steiner drew a simple sketch and spoke slowly

for me to take it down in writing: 'Sirius is the world-thought which Christ produces out of His Heart—therefore it is to be found within the earth.' He drew a curve to represent the earth and wrote on it 'Metabolism and fulfilment', as though the thoughts issuing from the Heart of Christ—that is from the sun—are sent through Sirius to the centre of the earth, where they obtain their fulfilment by means of metabolism.

Rudolf Steiner also spoke to me at that time about 'Creation out of nothing'. I noted down the following: the interior of the earth consists of gold, in which the highest hierarchy weaves. This interior became dark; and Christ departed from the sun and, as the Saviour, entered the golden centre of the earth. So is this new Creation a gift to the earth out of nothingness, out of the hollow space within the sun? For the solid earth—this body of the Gods—died at the moment of Golgotha. Has a new Holy Host now been sent down from the hollow space of the sun?

Rudolf Steiner answered: 'Not out of the hollowness, for that is "nothingness" isn't it, the opposite of "somethingness". The new Holy Host issues forth from the reality behind the emptiness.'

I: 'And will the mineral slag of the divine corpse become the vessel which begins to shine? Will all the mineral parts of the earth shine—or will a part of it remain behind, dark and unredeemed?'

Rudolf Steiner: 'All substance will be redeemed.' I then asked if the moons of the planets will also be redeemed and he answered: 'They are the refuse, yet they will also be redeemed—all substance will be redeemed.'

I: 'Is there any waste matter apart from the moons?' Rudolf Steiner: 'Yes, that will remain, but later on that too will be redeemed.' And then, with great seriousness: 'If everything were not to be redeemed, that would mean that Ahriman would have triumphed. All substance will gradually be redeemed after a very long time. If anything should be lost, then Ahriman would be the victor!'

I said: 'I am also looking for the relationship of the

Demiurgos (world-creator) to Ahriman.' He replied: 'The Demiurgos is, so to say, the young Ahriman.'

I: 'His building material is the mineral?' Rudolf Steiner nodded and I continued: 'And when the Demiurgos meets the Christ will this mineral then become the Holy Grail?'

Rudolf Steiner: 'Yes, that is correct.'

I then asked if the work done by W. Jordan about the Demiurgos was correct and Rudolf Steiner said: 'He has muddled it all up and made a hash of it'; and when I questioned him further as to where Jordan had acquired his knowledge, he said: 'He had an old initiation in the past.'

I went on: 'How is it possible that Zarathustra only spoke about the Sun-God and his opponent Ahriman and not about Lucifer?'

Rudolf Steiner answered in a tone of reproach: 'Because he did not deem it necessary.' But still I persisted: 'But Lucifer was also there, people must have known that!'

Rudolf Steiner: 'No, people only saw that part of the spirit which Zarathustra revealed to them.'

Rudolf Steiner then made a little drawing (which was later lost in East Germany) and he said: 'Lucifer's realm only stretches as far as the moon, but the teachings of Zarathustra reach to the sun sphere which is behind it.'

I had also noted down the question: 'What is the meaning of the Apocalyptic phrase: "When the moon became as blood"?' (Rev. 6,12).

Rudolf Steiner said in a most serious manner: 'That is a very deep saying which you have quoted!'

I: 'Does it mean that the dark moon slag begins to shine when the red light shines through it from within—or does a red glow shine from the visible sun towards the moon?'

Rudolf Steiner: 'The first statement is correct, not the second. The latter is incorrect. The moon is illuminated from within—you see, it is like an egg: the yolk is the moon through which shines the red light.'

In the new Uhland edition [of *Parsifal*] which has a foreword by Rudolf Steiner, an episode occurs which is not found anywhere else: the journey of the young Grail Queen Repanse, who is awaited in India as the wife of 'Fire-Fils' (Feirefis). She had embarked in Marseilles to take the emerald-green vessel, containing the Blood of Christ, the Grail, to the land of Priest-King John. When the Grail was carried ashore, the Grail Castle upon the heights lit up in flames to receive it.

I thought about this for a long while, and the 'Fire-Sons' of India who are to appear as magnificent figures at some future time, merged in my mind with the word Fire-Fils, which has the same meaning as Fire-Son. Feirefis is the dark brother of Parsifal, who is to succeed the latter as Grail King. Following on this thought I then enquired: 'Is Feirefis a Fire-Son?'

Rudolf Steiner: 'Yes, Feirefis is the Fire-Son.'

I: 'And Parsifal the Light-Son?'

Rudolf Steiner: 'Yes,' and then he said a few words which I wanted to write down, but could not quite manage, and so he dictated them to me: 'Parsifal is the Light-Son. He is the only incarnation of an important individual on earth.'

Thus: Parsifal, the Son of Light (= Abel), and Feirefis, the Son of Fire (= Cain) are the sons of God, divine brothers of man, whom he has lost—and whom Seth, the son of earth, must get back again. They meet us in three streams: the golden Arthurian knights or Michael/Sun knights; the silver knights, who guard the Grail through the holy wisdom of the moon; and the iron knights, who, in black armour and with the diamond lance, tarry in mortality until, conquering death, Christ with His light will penetrate the darkness.

I continued to question: 'Can one then believe that the Grail Castle is really present in the etheric world? "Travellers in the spiritual world" are forever telling us about it?'

Rudolf Steiner: 'Yes, it is so. It is really present.'

I: The Grail story, *The Chymical Wedding of Christian Rosenkreuz* and the Mithras-Cult are all aware of this Castle—is it the same sacred building that in the Bible is called "The New Jerusalem"?'

Rudolf Steiner: 'The New Jerusalem of which the Bible

speaks is the eternal, archetypal image of how it will be in the future. The Grail Castle is the image of how it is now in the spiritual world.'

I: 'Then it is really present spiritually?'

Rudolf Steiner: 'Yes, certainly—one can reach it when one is able, through occult training, to observe oneself from without—when one has become free from oneself. Then, from one's 42nd year onwards, one can trace one's life backwards to one's own pre-birthly condition. If one actually goes back to this moment, one is then within the Grail Castle.'

I: 'Then one experiences images of the Grail Castle in imagination?' Rudolf Steiner confirmed it. It was my last conversation with him.

Then came the leave-taking which at other times had been so gay: everyone full of smiles and talking of the next occasion. Now all were troubled. Herr Doctor shook hands with each one in silence; this time he did not say 'auf Wiedersehen'.

Frau Dr Steiner was driven in an open car to Breslau by Herr Grunelius. We took Dr Steiner somewhat later to the station. Frau von Moltke and several others were at the station so that they could see Rudolf Steiner once more.

When, after all these days of happiness were over, I sat again quietly at my writing table, the door was opened and our trustworthy old chauffeur came in and told me he had something important to say: the axle of the car had broken! For twelve days he had driven Dr Steiner without an accident and now, on the first drive after it, this had happened! And then he had something remarkable to tell me. He had had such rheumatism in his arm that he could no longer control the steering wheel. He had only waited for the conference to end and then he was going to ask for other employment. When, in great pain, he had driven Dr Steiner on the first day, the pain had lessened and then disappeared entirely. He remained as our chauffeur and the pains never returned.

Other members of the household spoke to me about how beautiful the time had been, and the old butler said: 'It could have gone on for another ten days and it would not have been too much for us.'

And Herr von Koschützki described his impressions of the Koberwitz Conference in a letter to the priests:

'If, in conclusion, I try to convey the total impression of the God-given gift of these Whitsuntide days, I must say to you that this can only be understood by someone who, once at least in his life has been madly in love. Now, raise this feeling up from the senses to the spirit, and you have my impression.'

Dr Günther Wachsmuth reported on Dr Steiner's journey from Koberwitz to Jena on 17 June 1924: 'It still lives vividly in my memory how during the journey, after looking back quietly and contemplatively on the conference for a while, he suddenly exclaimed in a strong, delighted tone: "Now we have accomplished that important work as well." Seldom have I found Rudolf Steiner so happily stirred and visibly delighted after an accomplished deed as on this occasion after the Agricultural Conference.'

Concluding notes by Count Adalbert Keyserlingk

If one looks back from the seventies of our century, the events of Koberwitz show us how developments since the twenties, and still more since the Second World War, have set us problems to which bio-dynamic agriculture offers practical solutions, and not only in the field of agriculture.

It seems to me as if the destiny of my father, Count Carl von Keyserlingk, is an image of what takes place when the earth receives an injection of new impulses at times of crisis.

He stands in the ranks of those who at such times unconditionally place themselves at the disposal of spiritual innovation.

After having engaged Dr Burgk to administer the experimental section of the properties of which he had charge, and ensuring that the latter obtained his Professorship in Giessen, he then turned his attention to the new ideas which Rudolf Steiner was able to give to agriculture. Professor Burgk, however, soon began—on account of these very ideas—to oppose my father in the business. The results of that have been described in these memoirs.

After crushing difficulties during the three years following the Agricultural Course, my father decided to give up his work with the firm and also in Koberwitz, in order to put the ideas of Rudolf Steiner quietly into practice on two of his own estates. Once again it was destiny which enabled him to buy these two estates with the help of Eliza von Moltke, through her family connections.

Burgk took over my father's position and bought himself an estate in North West Germany. But he destroyed what he had thus acquired by setting fire to it, and then committed suicide.

Such experiences and all the things my father had gone through with representatives from the big chemical fertiliser works, showed him what sort of opponents the young Biodynamic movement was up against. That did not run according to objective scientific rules! Nor was it possible either to convince impulsive young members of the seriousness of the situation, nor to talk to them about the plans for implementing new methods in agriculture in Germany which had been drawn up together with Günther Wachsmuth and Ita Wegman. My father, however, had gathered together a group of helpers, with whom he wanted to obtain the first practical results from these new methods. But even in that he met one disappointment after another.

Renewed opposition also arose within the family: inspite of a family fortune of millions, financial restrictions were imposed which hindered extensive conversion of the two estates of Sasterhausen and Raaben. On top of that came the difficulties within the Executive Council (*Vorstand*) of the Anthroposophical Society after Rudolf Steiner's death.

My father's heart was broken. He died of thrombosis on 29 December 1928 on a journey to Dornach, where he wanted to act as mediator. Along with his personal contribution, his contribution of land and money also came to an end.

We moved to Koberwitz four years before the Agricultural Course; and we moved out four years after it. Yet Koberwitz and Carl Keyserlingk were for Rudolf Steiner the place and person which offered the first vessel for a new agriculture, in which it could be carried out into the world. These first vessels broke so that something new could arise from their seeds.

PART II
Accounts by those who were present

Count Alexander von Keyserlingk

When my father returned to Estonia after a visit to Germany, he reported having visited his brother and other relatives in Silesia, who actually lived quite close to one another, but had very different views about the world and mankind. He had been very strongly warned by his brother not to enter the house of Carl Keyserlingk, because the family pursued very dubious and misguided paths. But when he paid Carl a visit nevertheless, everything made such a spiritual, as well as a fresh and modern impression on him, that the two of them soon became engaged in a lively conversation and found many things in common. The social attitude of Uncle Carl was something quite unusual for Germany in those days—I should, he said, definitely get to know it.

At that time I already had quite a lot of experience and I knew that the world was in a ferment. I had lived on the Baltic estates and in Tsarist Petersburg, had for years been in Eastern Russia, in Siberia and the Altai Mountains, experiencing the bloody start of the communist experiment as a member of the White Russian army, and had fought in the First World War. I had a burning interest in everything that was new. My very first conversation with uncle Carl showed me that my father had not exaggerated; and when uncle Carl found out that I had always worked as a farmer and had been intensively occupied with cattle-breeding, he asked me if I would stay with him and help him to start a pedigree herd on his estates.

I soon acquired a good overview of what was wanted and done there. I knew, too, that uncle Carl and Aunt Johanna were involved in anthroposophy. I already had a notion of what that involved, for I was a reader of the periodical *Der Türmer* ('The Watchman') in Estonia and had followed the arguments between Rittelmeyer and Lienhardt in it. I was well able to understand what Rittelmeyer wrote and therefore already had a positive attitude at the time I arrived in Koberwitz. I was put to work in the Experimental Section, in which Erhard Bartsch, who had just completed his studies, was also placed.

The firm Vom Rath, Schoeller and Skene included the great sugar factory in Klettendorf and 18 estates comprising 30,000 acres in total (7,500 hectars) of the best black earth. The estates supplied the factory with sugar beet. Uncle Carl belonged to the board of directors and was in charge of the whole of the agricultural section. 82 officials and employees were stationed at the head office in Klettendorf and over 1,000 worked in the fields. Every estate was run by one or two inspectors and the necessary assistants and trainees. Professor Burgk, a lecturer from Gießen University, was engaged by uncle Carl as the expert in charge, to run the farms as model estates, employing the most modern methods and producing the highest yields.

It is necessary to have a large herd of cattle to cultivate sugar beet, but the herd has to be replaced frequently due to its one-sided diet. It went against the grain for uncle Carl to buy 1,000 head of cattle every year anew, then for them to be slaughtered for beef; so my task was to build up a pedigree herd instead. We found a tract of land in East Prussia for rearing the young stock.

Everything was arranged according to the latest methods. A light railway connected the estates to the sugar factory and the beet was transported by an elevated track directly from the fields to the processing plant. The rotation of crops was planned 10 or 12 years in advance and the whole concern was so large that the experimental section was kept busy negotiating plans with powerful inspectors, laying out testing plots and supervising all that the work entailed.

Uncle Carl had, of course, to be responsible for all the needs and interests of the factory. Afterwards, however, he had to take on the running of the firm as a whole, which meant, because of his social concern for his workers and the fact that he was an anthroposophist, that he was frequently engaged in internal squabbles. On one occasion his son

Wolfgang asked him, on a car journey, why he always worked longer than the others and never allowed himself rest or enjoyment. To this he answered very seriously: 'I am responsible for the well-being of more that three thousand people. Their position has to be safeguarded, and that can only happen if the firm is strong.' At that time he built workers' houses and, to the annoyance of some of the gentlemen, created a 'social section' which was under his sole direction. Every employee could come to him—without having to ask his direct superior—and express his fears, wishes or suggestions. At that time, because of the catastrophic shortage of jobs, it was dangerous to make any complaints. In order to make sure there were no reprisals, a monthly discussion was arranged, to which every estate sent two workers' delegates who could voice their concerns in strict confidence. Shop committees did not exist at that time. This arrangement provided the workforce with great assurance and there were never any strikes.

There were three main political camps at the time: Nationalists, Democrats and Communists. If an applicant coming to seek employment pretended to be staunchly nationalistic, in order to create a good impression, Uncle Carl was likely to tell him in a serious but friendly manner: 'Your political views are your own affair. They do not concern us. The only important thing as far as I am concerned is that you carry out your work well and are a decent human being.'

During this inflationary period Uncle Carl introduced a quite new way to pay wages. Although currency lost its value every day, wages were generally paid weekly. Large firms were allowed to mint their own coinage—whose value was a little more constant—which could alleviate the greatest hardship. Huge piles of money lay in their cellars. But uncle Carl had a better idea than that, one which certainly made a lot of work for the office staff, but was of great benefit to the employees. From one day to the next was calculated what each had earned—not in German Marks, but in pounds of rye! It was of no advantage to the firm, but the workers got the equivalent value of their pounds of rye on pay-day. If they then changed it immediately, there would be no loss. People could even save without losing out! But of course this led to great difficulties with the firm's financial directors.

Uncle Carl succeeded in converting the family business into a limited company—whose board of directors included people who, either openly or in a hidden way, made things difficult for him from the very start, because of his anthroposophy. The workers, however, already knew who had helped them, and made this abundantly clear on a dramatic occasion when an attempt was made to dismiss uncle Carl from his post.

Many of our neighbours called uncle Carl 'The Red Count', which was not exactly a term of endearment, but fitted quite well with the title 'The Iron Count', which Rudolf Steiner gave him.

Uncle Carl was exemplary in the way he kept in mind both the firm as a whole and also the welfare of every single herd-boy on the estate. On our tours round the estate he could suddenly call a halt and ask someone: 'Is anything bothering you?' He would learn a lot in this way—things which were just as important to him as some of the larger matters.

This was a time of the worst revolutionary excesses. Sometimes one or other of the workers would fail to greet uncle Carl as he was walking across a yard. Anyone else in authority would have reprimanded the man immediately, but uncle Carl would just look at him in amazement and walk on. On the next occasion the man would certainly greet him.

If we young people had been invited to celebrations in the villages, we were not allowed to use either the coach or the car, but had to walk or go by bicycle, for uncle Carl thought it was unfriendly or tactless to drive past other people, and thus force them to move aside and possibly get splashed.

At first I lived in the mansion and was soon also given charge of the forestry and game department. We had six or eight foresters for the copses—there were no woods on the valuable Silesian soil. I introduced ducks on the estate, laid hedges, made watering places and organised the shoots of pheasant and hares. And I also had my falcons.

There were always guests at table. I think we were never less than 15 persons. The first anthroposophists I met there were Herr and Frau Scheer, with whom I had many discussions. Also various priests of the Christian Community, including Rudolf Meyer. Koschützki and his wife lived there, as well as Rittelmeyer too for a long time. The latter was very humorous, but even when he said something in fun, it stuck in one's mind afterwards. On one occasion he had just arrived from Berlin. As we took our places at table everyone was hoping for some flashes of wit from him and asked him what he had been doing there. Then he said with a very serious face: 'I experienced something very important. I went to the Busch Circus.' And as everyone gazed at him in disbelief and amazement he continued: 'One only has to study the concentration of the ring-master and the artistes! If anthroposophists would only work with the same concentration they would be able to achieve things of unbelievable benefit to the world!'

Her Excellency Frau von Moltke was often present. I was annoyed by her rather domineering presence, and the fact that she always sat in Aunt Johanna's seat as a matter of course. But a talk with her was always interesting. Especially impressive for me was her story of the egg.

One evening she sat at a little table darning. In front of her was a darning egg, apparently made of glass. I took hold of it, but she called out: 'Put it down immediately! It will certainly dissolve again if it is handled by such a prosaic person!'

And then she told the following story:

'I had once been invited to a Spiritualist meeting in Berlin as an objective observer. A medium was there, a poor woman for whom I felt pity, as she was quite exhausted and was being exploited by the people in charge of her. At the end of the séance she asked for a glass of water. However, she did not take a drink, but took it in her hand and went into a trance again. She lifted the glass up high and suddenly the water began to rotate. It continued thus for some time and then came to a standstill and was solid. The lady woke up, came towards me and tipped this egg into my hand.

I then asked her for her address and visited her a few days later in a suburb. She met me with a basket of washing which she was about to hang up, beamed with pleasure and said: "At last there is someone who takes a personal interest in me—I noticed that straight away!" In saying this she reached up into the air and plucked a couple of roses which she presented to me. I took the roses home with me and they were just like any other roses.'

I then asked Frau von Moltke what the egg was actually made of, and she told me: 'I once showed it to Rudolf Steiner and asked him if it was glass or crystal. He took it in his hand and answered: "It is neither glass nor crystal, but magnetised water." Once I knew that, I occasionally led water diviners a merry dance, by hiding it under the carpet and getting them to dowse for water. The divining rod always reacted.'

Her Excellency also mentioned that she was able to cure headaches with it. She never allowed me to touch it again.

When I think back on the discussions in Koberwitz and the mood that prevailed among us, I realise that we youngsters were just as critical, nonchalant and at times provocative in our attitude towards the 'older anthroposophists' as the younger generation is today. But that does not mean that we did not concern ourselves with all sorts of problems in just as serious-minded, if perhaps less emotional way. A little anecdote illustrates this very well:

It happened during the Christmas Conference in Dornach. The verses for the laying of the Foundation Stone were the most impressive event. On one occasion they lasted well into the night and whilst they were being spoken a raging storm broke out. When Rudolf Steiner was asked why this had happened he replied: 'The elemental beings were afraid that they were not being included, and that accounts for the last words of the verse as we know it.'

Afterwards, as we stood in front of the Goetheanum in the night, I noticed how quiet nature was, although the storm had raged shortly before.

How did the Agricultural Course come about? Vögele was

no longer in Koberwitz at that time, but Dreidex was there and Erhard Bartsch, whom I helped study for a diploma. I was occasionally with uncle Carl at Herr Stegemann's, and it was said there that Rudolf Steiner was to be asked to arrange a farmers' conference. Stegemann already worked according to the advice of Rudolf Steiner at that time—however he did not yet have the Preparations.

All of us were aware of the fact that the use of chemical fertilisers was making the soil's future, and that of man and animal, look ever bleaker, even though higher yields at first disguised the true state of affairs. Owing to increased numbers of nematode worms, the amount of sugar beet one could grow was constantly dwindling—and there was no remedy for this. When we saw from Stegemann's results how well Rudolf Steiner understood this problem, we wanted to learn more about it. We did not know that it would be a whole course of lectures, nor did we guess what wide perspectives Rudolf Steiner would bring to bear on the subject. We only thought he would give us a few instructions to counteract the destruction of the soil's fertility and the diminishing quality of some varieties of fruit. Uncle Carl and aunt Johanna must have had discussions with him about it, however, before I came to Koberwitz.

On the occasion of a journey I made to Dornach, uncle Carl had given me instructions to ask the Doctor for a date for a conference about the new agricultural methods. When I arrived in Dornach I went straight to the workshop and told Frau Dr Steiner that I would like to have a word with Dr Steiner. I did not have long to wait. He appeared and I said my piece. He at once said: 'Yes, I will come to Breslau and will give lectures about agriculture there.' But I told him that was not sufficient—'I was not instructed to ask if you would come, but when you are coming!' Then Doctor Steiner smiled, took out his notebook, turned over the pages and then announced: 'Tell your uncle that I will come to you at Whitsuntide.'

The planning started in spring. In order to stage a big conference, great preparations and financial means are necessary. Only about 120 people were allowed to attend the course itself, but the daily number of guests was far greater. The firm's transport was used and the whole house was full. Servants, chauffeurs and gardeners were put in uniform and all were given their duties. The midday meal was usually eaten standing, because the dining hall had to be cleared for the lectures. On all sides there was an expectant coming and going, and when Rudolf Steiner made his appearance a wonderful atmosphere reigned over everything and everyone.

A posse of young people was busy on his behalf. They carried out his instructions, kept watch and relieved each other, just like adjutants of the General Staff.

I had heard that Rudolf Steiner often worked for hours into the night after the lectures, then walked through the silent, spacious house, descended the stairs, opened the front door and handed his letters for posting to Herr Grunelius, who was waiting in his car to catch the early postal train with them. So I told Rudolf Steiner that, since he went to bed so late and was first up in the morning, he ought not to have to walk through the whole house as well in the middle of the night. I would in future collect his letters. He said to me: 'But you too need your sleep!' I told him it was more important for him to get some sleep. He answered with neither a yea nor a nay, and so I knocked on his door at about half past three in the morning. On his 'Come in', I entered the room and saw him sitting up in bed writing. He only said: 'It will still take a few moments.' On returning some minutes later he had finished the letters and gave them to me. I did the same thing every morning.

As I was standing by my aviary very early one morning—I was doing an experiment at that time to see if I could acclimatise budgerigars—Rudolf Steiner came out of the house and stood beside me watching the hundred or so creatures; then he said: 'You can also breed blue ones.' At that time there were no blue budgerigars—only greenish-yellow varieties were known.

During the course the morning lecture always lasted till

lunch-time, then there was a break for a snack during which sausages were served, followed by a discussion till 3.00 pm. In the afternoon one could walk in the park, or the stables could be visited. In the evening everyone drove to Breslau for the lecture and only on our return was the evening meal served. Although this programme put a heavier burden on Rudolf Steiner than any of us could have borne, yet it seemed to me that he looked better and happier day by day.

At the end of the course, the 'Experimental Circle' was founded—and that was when certain difficulties began to arise. There was a group of enthusiastic young farmers who wanted, without delay, to blazon about all they had heard. Uncle Carl, however, had made a strong plea for us to wait with the news until we had some practical results to show. He warned us that every word which was spoken on the subject could be twisted round and would then constitute a danger for these new methods of agriculture.

He later told me with great sadness that he had been invited to a meeting of the J.G. Farben fertiliser company, where news of the new methods of agriculture had been spread abroad, and they had been very 'interested' in it. They had asked him if he could provide them with a copy of the Agricultural Course (which had numbered copies and were only given to members of the 'Experimental Circle'). Finally he had been offered a good round sum for it. When he still did not accept this offer, they plied him with questions. While he was talking to them he noticed that the ash-tray in front of him on the table was of a very peculiar shape and he discovered that it contained recording equipment. This drastic experience in connection with one of Germany's biggest artificial fertiliser firms confirmed him in his views. He had, however, been unable to convince friends that everything, including relationships within the firm, were not nearly as simple as they were believed to be. For uncle Carl could not at that time simply separate one of the family estates from the rest to supply the sugar beet factory with beet grown by the new methods, as some people expected him to do. We certainly made immediate tests on certain fields and plots of land—but, in order to apply the Koberwitz Course in practical life, one would need to have land which could be managed as one pleased. Therefore uncle Carl asked to be paid off, and bought two estates of his own land without incurring debt. He had to leave the firm if he wished to work further along the lines of the Koberwitz impulse and for anthroposophy.

From the very beginning uncle Carl wanted to incorporate into his work [anthroposophical] social ideas which were very dear to his heart. Looking towards the future, Rudolf Steiner had given them at the same time as the new agricultural methods. This, however, would only have been possible if everyone who had participated in the course had worked together—but some of them immediately wished to make themselves independent of the rest, in a head-strong kind of way. These were enthusiasts who accepted the methods as far as manuring was concerned, but neglected the esoteric and social side, which can only be realised through community.

Uncle Carl clearly saw this. He said to me: 'However well it functions economically, Germany's future will in no sense be served by failing to carry out what Rudolf Steiner impressed so deeply on our hearts—the integration of esoteric and social aspects into biodynamic methods of agriculture! Without that it might exist for a while and yet would not have real future potential. However hard one might strive to keep alive a half-baked affair, it would remain without effect upon the world until all the impulses which Dr Steiner gave us in the Agricultural Course have been taken up.'

Paula Eckardt

Koberwitz—what a wonderful time that was! What experiences one had there! On Sunday mornings the old fellow Fender always stood there at his service-room window and looked out to see how many would turn up, so that he could set the table.

That was during the bad times; people were glad to get out into the country. There was a great muddle and fuss in Germany—but with us in Koberwitz it was like paradise, an island of peace and order—whilst all around was tumult and uprisings! Things had broken up in Saxony, the Reds were in Hamburg, the Browns in Munich, and we had inflation. Well, we had all sorts turning up: the friends of Count Wolfgang, the anthroposophists and all the other interesting people.

Excellence von Moltke, she was an energetic one! She came to the door, knocked with her stick and said: 'Here I am again!'

And our Herr Count, she was so fond of him! He often had to go up to her room in the morning, even when she lay in bed, and then she always had something important to discuss with him. She was very angry with the Kaiser—that had something to do with her husband—but she got on very well with Prince Sigismund—it was not his fault, of course! Those were lovely people—Princess Sigismund always gave me a warm hug—and the Prince was very unassuming; he once asked me very bashfully if I would iron his collar for him, saying he had washed it already. They went to Africa later—or was it America? At any rate—when she had her first child, they were so far into the interior of the country that an African had to attend on her. I was told all that by the Princess. They were certainly not demanding, neither was her Excellency, Frau von Moltke. She had such wiry hair! I had to go with her up to her room in the evenings to plait it. She didn't give herself airs—one could just be one's natural self in her company. She was more like a soldier!

And Doctor Steiner himself—I remember him very well! He was very calm. If he were left in peace he would work in his room and never emerge. Modest! Not at all like a great man. His wife was more lively.

The other maidservants in the house saw the Doctor less often that I did, because I had sole charge of him. He always greeted me in such a friendly manner—not at all high and mighty! He was a lamb alongside the other anthroposophists! Every morning I took him his tray: coffee especially, toast and honey, nothing out of the ordinary. Then he breakfasted with Frau Doctor—often it was the only quiet moment he had all day.

All of them had breakfast in their rooms at the same time. Our meals were always very punctual—it had never been otherwise in this big household! We housemaids often never saw one another in the mornings until each had completed her section. And don't think that the Countess didn't have her finger on the pulse—even though she studied ever so much, she knew what was what! She had discussed the timetable and how the food was to be prepared and how many people there were and what was needed. And she was so generous—gave a lot away, especially to anthroposophists. That's how it was! And that set some people against her.

She sometimes came and said: 'Now I do not want to be disturbed, folk!' And she would disappear for two or three days. Then she had her meals sent up to her. And if there were guests present—and there were always some there—they felt quite at home with Fräulein Pietsch, our housekeeper.

Or she might say: 'Folk, take the day off and go to Breslau, there's a good film on.' And when I said: 'But I've just sprinkled all the shirts, Countess!' Then she would say: 'You can still iron them tomorrow.' And so we drove off with the coachman and the old man to accompany us.

In September she sent all the guests away. We also had to have our holidays. She was very precise about free time—and the household ran without a hitch—except for the Course, you could say. Then it was a case of 'all hands on deck!'

And as for the Count—he was more than kind! Once I had nodded off when the coffee was to have been served. He noticed that and would not allow anyone to wake me up. He had a fine soul, if you like!

Well, and as for Count Wolfgang—he was not so much in favour of anthroposophy. He made himself scarce and did not want to get involved—but the Countess herself, she lived for nothing else.

And when the Course came about—Oh dear me! What a lot of preparation was needed! All the rooms had to be moved around. Then flowers had to be put everywhere, all over the house.

Dr Steiner was always put up in the Countess's bedroom and drawing room, and Frau Doctor in Frau von Moltke's room, for her Excellency never stayed with us during the Course. Frau Dr Steiner also gave me a book: *How to Attain Knowledge of Higher Worlds*. The Russians took that from me—and it had Doctor Steiner's signature in it! I could buy myself another copy now, but I do not want to take it back with me [Paula Eckardt lived in East Germany] and anyway, my old eyes can't read anything new—it's too late to learn anything I don't know by now.

During the Course a lot of people came over from Breslau for breakfast. And at nine o'clock the lecture started. After the doors had been shut we worked away like the Fire Brigade. Everything was brought to make the meal for all those people. During break-time there were sausages and slices of bread—we had prepared mountains of it—and whatever anyone wanted to drink. Herr Doctor was sometimes surprised and laughed at the speed with which the platters were emptied. Not everyone went to all the lectures. There were also the pretty eurythmists—they always had their escorts when they went for a stroll in the park! They stayed in Breslau and the young people stayed with us in the Forestry house. They also had a good time—after all the talking and lectures in the Mansion.

All went off to Breslau in the afternoons, and Herr Doctor went too, a bit later. And when they returned in the evening there was another hot meal. All went without a hitch!

During the Course, and also before, I got to know a great many wonderful people: Pastor Rittelmeyer and Pastor Geyer, Herr Bock, Herr Vögele and Rector Bartsch, the young Wachsmuth and whatever others there were. The Countess always said to them: 'Come and pay us a visit.' So they came and wanted to stay a day or two—and after a couple of weeks, or a couple of months they were still there!

But the Course—what shall I say—that was a mighty affair. I would be able to recognise Herr Doctor Steiner today among thousands! He said goodbye to each one of us and thanked us for the pains we had taken. Yes! That was Whitsuntide 1924 and in 1925 Herr Doctor died. Would you believe it!

Luise von Zastrow

Crowds of participants came early in the morning to the Agricultural Course in Koberwitz, streaming in long queues past Rudolf Steiner and the Count, who stood waiting in the open by the door. The guests then congregated in the large hall, where a lecture with discussions about agricultural questions took place every morning in the adjoining dining room. During the break the guests were regaled in a most generous fashion by the Keyserlingks. Many sought refreshment in the park with its picturesque lakes, and thought that they had never before experienced a more radiant Whitsuntide weather.

A daily changing circle of participants was invited for lunch by the Keyserlingks after the morning activities had finished. The lecture hall was converted into a dining room with great haste. A festive meal was laid out on a table for 12 people. The rest of the guests and members of the household ate in the adjoining hall.

As I was at that time the part-owner of an estate in the Glatz district, and also a colleague of the Count and Countess Keyserlingk, I had the good fortune of being able to take part in the Agricultural Course and was invited once or twice to a meal in this intimate circle.

As though attracted by an unseen force, the eyes of all the guests turned in the direction of the door when Rudolf Steiner entered the dining room. With upright stance and long black coat the Doctor stepped inside and greeted everyone individually with distinguished and calm dignity and kindness.

I should now like to describe once more my recollections of the conversations at table. When I was present for the first time, Countess Keyserlingk apologised for having forgotten to provide a glass of water for the speaker, Herr Doctor:

'That does not matter,' replied Rudolf Steiner, 'I never

drink anything during a lecture.' After a short pause he added that he had become accustomed to his audience drinking when he was lecturing to the Workers' Educational Association: 'And when I was speaking about Heraclitus—"everything flows"—one of the workers upset his beer mug and then everything really did flow!'

At mealtimes, throughout his stay at Koberwitz, one could observe how Rudolf Steiner improved visibly, even though he was giving at least two or three lectures every day and holding discussion classes. On top of this he was occupied all day long in talking to people and giving them advice.

The priest, Rudolf von Koschützki, was once prompted to tell about his railway accident, from which he miraculously escaped with injuries after having been wedged under train-wreckage all night without losing consciousness.

He was sitting at the table between Günther Wachsmuth and me, and thus I was able to study from close at hand the deeply serious face of Rudolf Steiner, who looked at the speaker in silence the whole time. One could see by his countenance that he took in more of what had happened than the description of this event could express.

Herr von Koschützki described in his books how this accident was like eternal night.

Rudolf Steiner was surprised that none of the members asked him questions. It was his state of health which up until then had made people reluctant to trouble him.

But now many people came forward for an interview. I too asked to be allowed to speak to him, but I vowed not to ask him for anything personal.

Now I was able to look into Rudolf Steiner's eyes from a very close distance and experience their golden sheen. To begin with I could tell him of my great debt of gratitude to him.

On the following evening, during a karma lecture, Rudolf Steiner mentioned how a feeling of gratitude opened doorways to the spiritual world.

In addition I asked Rudolf Steiner about the writings of Countess Keyserlingk, which appeared to me of such importance and to be so spiritually stimulating that I had made it one of my life's tasks to help her to arrange them and have them distributed. But I would do so only if it had his approval.

Rudolf Steiner replied: 'Yes, the papers of the Countess are interesting and would be stimulating for a certain circle of people.'

When the interview came to an end Rudolf Steiner came to the door with me and uttered a thrice-spoken 'Auf Wiedersehen'.

Rudolf Steiner mentioned what I had proposed to Countess Keyserlingk on several occasions and supported my suggestion that her works should be made available to a wider public.

After this conversation the Countess and I carried out many extensive, interrelated activities together.

After the address given to the young people, the hour of departure approached. The Countess quickly went to fetch the visitors' book.

She asked Rudolf Steiner to write something in it and this he did, as in 1922, but this time without any preparation:

> With love to the house of Koberwitz,
> The seat of good anthroposophists,
> We came to search anew
> For the hearts both faithful and true,
> The active spirit to rouse
> Which lives within this house
> By the love which here we found
> We are to each other bound.
>   With most heartfelt thanks
>   Rudolf Steiner

After that he handed the visitors' book to Frau Marie Steiner and said: 'So—now you add a meaningful thought to it.' 'No,' she answered, 'I won't do that, I shall only add my name, I prefer to put myself in your shadow.'

All the inhabitants of the house stood in the hall and at the door during the leave-taking. Rudolf Steiner went up to each one individually, including the man servant and the maid-servant, and held out his hand to them without saying anything. His thrice-spoken, friendly 'Aufwiedersehen' resounded no more. We were touched by the breath of departure.

The Whitsuntide gathering in Koberwitz had come to an end.

Günther Sponholz

Of the actors belonging to the Kugelmann Group, who took part in the performance in Breslau, I am the only one left alive today. As, in my position as tour-manager, I also had to organise any guest performances, I first of all travelled to Breslau. There I had my first meeting with Marie Steiner, who happened to be at the dress rehearsal of a eurythmy performance at the Lobe Theatre.

She had been fixed up with a small reading and control-desk between the rows of seats in the completely darkened auditorium. From this position she kept an eye on the eurythmy proceedings on the stage and gave her corrections.

She was very interested in my proposal that we should show Rudolf Steiner some of our work. She asked some very animated questions that were concluded by a decision to discuss everything first with Rudolf Steiner. I could have my answer, however, the same day, after Rudolf Steiner's evening lecture.

The performance had to be fitted into the Agricultural Conference programme at the last minute.

Rudolf Steiner gave his agreement, but said I should first drive to Count Keyserlingk to get *his* approval. I arrived early at Koberwitz with the many other people who wished to hear the lectures. Count Keyserlingk stood at the door to greet the guests.

I delivered my request straight away and assured him that Rudolf Steiner would also like it to take place. Count Keyserlingk answered: 'Yes, bring the *Iphigenie* Group to Breslau, we shall do all we can if Rudolf Steiner wants it.' I replied: 'But it will cost a lot of money to bring the entire company from Rostock to Breslau.' 'Yes, I will pay for it if Rudolf Steiner wants it.'

So I was able to arrange it and bring about the change of

programme. On 15 June 1924 Goethe's *Iphigenie* was performed on a large, but simple curtained stage in the factory.

Georg Kugelmann gave some introductory words about the Ensemble's beginning, about the work on speech and the further aims of the group.

During the performance—in the interval—Rudolf and Marie Steiner came onto the stage. They wanted to see Kugelmann and be introduced to the Ensemble.

Marie Steiner spoke first. She thanked Kugelmann with the following words: 'At last we see something achieved out of all the stimulus that Rudolf Steiner has given!'

Rudolf Steiner held out his hand to each one and said: 'I have been asked by many different people to give a course on dramatic art—and Herr Haas-Berkow has also come to me here in Breslau for that purpose, to ask if I would hold the course this summer. If, however, your duties do not allow you to get away, I shall not hold the course until next year.' We naturally accepted with great pleasure. The aim of all our work could perhaps be best summed up as follows: to work uncompromisingly on the basis of spiritual scientific methods. There were however two obstacles to this: the professional theatre and the amateur groups. Kugelmann explained this situation to Rudolf Steiner, who responded as follows: 'The professional theatre will develop in the future in an intellectual, Ahrimanic way. Not very much more can be hoped for. Amateur dramatics, on the other hand, will in many cases by led by strongly Luciferic personalities. Those are the powers from the past which are of no further significance today. The only thing of any value is what issues from the ego-forces of spiritual-scientific methods.

The Drama Course then took place in September 1924, on which occasion Rudolf Steiner allowed the Breslau *Iphigenie* to be performed on the workshop stage in Dornach.

Lutz Engel

The train steamed slowly across the broad plain which lay between Breslau on the Oder to the West and the Sudeten mountains to the East. It was the evening of Whit Saturday, 7 June 1924—a brilliant summer's day. There were a great number of passengers on the train that day—we were all on our way to Koberwitz where the Agricultural Course was to begin.

My thoughts wandered off into the past. How much I had experienced during these last years! After months in a military hospital I had come to Breslau in the last year of the war. I had got to know some good friends, met my wife and had started my medical studies in 1919. 'You can become officially engaged after you have passed your "Physicum" [Doctor's preclinical examination]' said my father. So there was no delay. I passed my exam in the autumn, the engagement was celebrated in October, we were married in the summer of 1920 and moved to Heidelberg. I entered for my state exams the following summer, which were concluded towards the end of the year, and a month later I obtained my Doctor's degree.

At the end of January we returned to Breslau, where my practical clinical training was to start. Our son Peter was born in May.

The start of a private practice was to follow. But after having rapidly completed my studies, passed my doctor's exams and founded a family in barely three years, events had started to run away with themselves.

It was one day in the spring of 1922 that I met my friend Werner Löwenfeld, who had a book tightly wedged beneath his arm. He was an odd chap, a sculptor, who could hardly get down to any modelling because of his excess of ideas. We shared many interests, especially pedagogical problems, and studied in a circle of friends such books as *Schule und*

Jugendkultur ('School and the Education of Youth'). When I asked him about the book he had with him, he replied in his enigmatic fashion: 'This is written by a man with more insight than Bernard Shaw!' Of course, I simply had to read this book—it was Rudolf Steiner's: *Education of the Child in the Light of Anthroposophy*.

The die had been cast. I knew that I had been granted the great good fortune to have found my teacher and discovered the world of my longing.

Rudolf Steiner came to Breslau at that time and gave a public lecture there. Thus I saw him for the first time the length of a lecture hall away. I cannot say that this first physical encounter made any particular impression on me. I did not need that.

A lively time of lecturing and preparation started in the town. The Christian Community was to be founded. I had hardly been interested in religious questions before that time—that was to change now. But I found myself in a similar position with regard to medicine—for during my hectically short study I had had no chance to devote myself to wider problems of medicine. I found psychiatry mysterious and fascinating, and distrusted the teachings of pathology and therapy. That altered dramatically after I met anthoposophy, for there I discovered a knowledge of therapy which my superiors and the head doctors in the hospital did not know about. Yet my attempts to talk with them about it went sadly astray!

And so October 1922 arrived, and with that my doctor's licence. Then I learned that a 'medical week' was to be arranged in Stuttgart, at which, besides anthroposophical doctors, Rudolf Steiner was also going to speak. I absolutely had to be there! All that was lacking was money for the journey. I therefore decided to go to see Rector Bartsch, the leader of the Breslau branch of the Anthroposophical Society.

I began as follows: 'Herr Bartsch, I would like to become a member of the Anthroposophical Society.' 'Very good,' said the latter, 'then you will first have to be a probationary member for one year. Membership can follow after that.' 'In that case, I would very much like to travel to Stuttgart to the "Doctors' Week",' I said, 'but I have no money for the journey.' 'That works out very well,' said Herr Bartsch, 'I am the executor of a fund which is dwindling for no other reason than inflation—take the money for that.' I did so, and now I was able, along with other medical lectures, to hear Rudolf Steiner, and even to shake hands with him at a tea-party at the conclusion of the course.

In the following year the Christmas Conference took place in Dornach. I did not participate in that, but heard a lot about it from my friend, Gerhard Suchantke, who was a student at that time. During the Conference a circle of young doctors and medical students was formed, to whom a more esoteric training was given for their professional practice.

And now, at Easter 1924, another Conference was to take place at which this circle was to be extended and carried further.

I felt that I had to travel to Dornach this time and there, in addition to the lectures for members, I took part in special medical events given to us by Rudolf Steiner in the Glass Studio. That was a very small circle of young people who were allowed to sit at Rudolf Steiner's feet and listen to him.

In the meantime—in November 1923—I had rented two rooms in Breslau, put up a doctor's nameplate and started my practice. It was the first anthroposophical medical practice in the town and, as there were many members of the Society there, it was not surprising that I had a full waiting room already on the first morning.

However—these anthroposophical patients! What completely new symptoms I encountered, of which I had never previously heard and which were not to be found in any medical handbook! I felt rather helpless in the face of this onslaught. But at Whitsuntide Rudolf Steiner was to come to give lectures to members, and the Agricultural Course was to be held in Koberwitz. How good it would be if I could present at least some of my difficult cases to him! Yet, Count Keyserlingk, who felt responsible for the conference, had made it clear to everyone that he would ensure that no one was to approach Rudolf Steiner with personal requests, for the latter was not well, and far too busy to cope with personal auditions alongside his lectures. And yet—my patients!

I did not dare approach Rudolf Steiner directly, but it could scarcely do harm, I thought, to write to his wife. The letter was posted. The fact that I did not receive an answer did not bother me further.

I pondered upon all that as the train rolled on towards Koberwitz. We got out at the small station and made our way to the mansion. Count Keyserlingk was at the door introducing Rudolf Steiner to those who were attending the course. Rudolf Steiner caught sight of me, beckoned me to him and said: 'Frau Dr Steiner is suffering from a weakness of the vocal chords and has to give a Speech course. Can you obtain some Pyrites 3x for me?' It was not such a simple matter at that time, for apothecaries had no remedies from International Laboratories Ltd in stock—and the post took a very long time. But I had just ordered this remedy in Stuttgart and it had arrived that morning. How glad I was to be able to be of use so promptly! Looking in my direction he continued: 'I heard that you wanted to present some of your patients to me, but that Count Keyserlingk has raised some difficulties. Well, if the Count will not allow it, I shall simply come to you in Breslau.'

'There is no question of that, Herr Doctor,' I called out, 'Count Keyserlingk only wants to protect you from overstraining yourself! If you explain to him that you would like to see my patients, I am sure he would agree and have it arranged!' And that is how it happened. A consultation was fixed for the following Wednesday afternoon in Koberwitz.

Yet the news of it quickly spread among those attending the course and many a one would have liked to have had some personal advice from Rudolf Steiner. So it came about that suddenly a kind of epidemic broke out. People came to me, explained very seriously that they were not feeling at all well and that only Rudolf Steiner could be of help to them. What should I do? Had I, a very young member, the right to prevent it? A second consultation-hour had to be arranged and it was only with difficulty that I was able to limit the number of patients to some twenty or so.

This is not the place to report on the medical consultations, but I would like briefly to mention something of the human aspect.

There was one tall blonde girl in her late twenties who had been in a sanatorium for pulmonary tuberculosis. She was better now, but had become hoarse, so that one could hardly understand what she said. Rudolf Steiner said that it was caused by a weakness of the etheric larynx. He prescribed a remedy and advised the patient to read aloud good pieces of poetry, such as Philia, Astrid and Luna's speeches from the seventh scene of the first Mystery Drama.

There was a fairly young farmer who presented something unusual: he could speak without difficulty but when he tried to say 'Mystery of Golgotha' he could only stutter My... My... My... with the greatest difficulty. Rudolf Steiner recommended medication and curative eurythmy.

Then there was a five-year-old boy. He had been born with one eye the size of a pea, which did not grow any bigger. Rudolf Steiner asked the mother what had happened during a certain month of her pregnancy. She recollected—she had had an argument with her mother in the kitchen. Mama had grabbed a pot and thrown it at her head, hitting her eye. In this case a very complicated treatment was prescribed with injections.

I brought him my two-year-old son, who bore traces of rickets, a slight deformation of the upper thigh. The treatment which Rudolf Steiner prescribed soon led to a complete recovery. While I was holding the child in my arms I noticed how Rudolf Steiner touched its head with his fingers. I thought he wanted to discover if the child's fontanelle was still open. I saw how he made certain signs with his hand on the child's head during which his glance was turned completely inwards. I was amazed. It looked as though he was giving the boy his blessing.

To one lady patient Rudolf Steiner had promised a personal meditation. I reminded him about this. He took his notebook out of his pocket and said to me: 'Write it down.' I got ready, then Rudolf Steiner spoke the meditation with a steady voice while writing it down. I wrote it too and later gave the paper to the lady. She took it up eagerly and, glancing down at it, called out: 'Oh, it is even in Dr Steiner's own handwriting.' Strangely, whilst I was writing down the meditation my hand-writing had become like that of Rudolf Steiner.

When the last patient had been attended to, Rudolf Steiner stood up. I thanked him and said that I also had a personal request to make—not to tell him that I was unwell, but that I nevertheless needed his help. He took up his notebook again, drew a simple sketch in it which he explained to me, then wrote the words of a meditation and gave me the sheet of paper.

And now I should like to recount what it was like when we were invited by Count and Countess Keyserlingk to have lunch together with Rudolf and Marie Steiner in Koberwitz.

Rudolf Steiner had just been speaking about the cosmic aspect which should be considered in relation to plants, certain trees and the control of pests. Herr and Frau F. were sitting with us at table and Herr F. said that he had something to contribute to the table-talk. So he began to discourse at large—and to brag a little: he had a connection to a professor of astronomy and would see to it that we had access to an observatory which would provide us with the necessary astronomical data. Rudolf Steiner listened to him at first without saying anything, then he interrupted the flow of words to say that he had a story to tell: Serenissimus had once visited his own national observatory. The astronomer there took the greatest care to explain everything as simply as possible to the Father of the country. Serenissimus strained his ears to listen. Then he shook his head: 'Wonderful, quite magnificent my dear Professor—but what is most amazing is that you even know the names of the stars!'

General laughter, the tension dissipated. Then I intervened: 'Herr Doctor, I'm on the side of Serenissimus in this case—I also think that to find the real names of the stars is the greatest achievement.' I was met by a rather astonished look.

Some months later, in Dornach, Rudolf Steiner was speaking to a circle of priests and doctors. It was at Michaelmas. He again told the joke about Serenissimus. Again there was general laughter. But then he said: 'Dear Friends, you need not laugh about Serenissimus. He was right—to know the true names of the stars is something great.'

I was sitting, as I like to do, in the front row and felt as if a surge of blood and warmth broke over me.

In Koberwitz at that time Herr F. consistently held forth and dominated the conversation at mealtimes. 'Herr Doctor,' he said, 'a method has been discovered for hardening wood, so that it becomes as solid as iron!'

'And I,' replied Rudolf Steiner, 'have discovered a method of softening wood so that it can be kneaded.' But Herr F. did not give up: 'I have heard,' he said 'that you have been doing experiments with peat to change it into a useful material for making into clothes. How far have you got with these experiments?'

'Well,' said Rudolf Steiner, 'we can already produce a material of the quality of hessian.'

Herr F.: 'When you have produced a useful material I will have a jacket made of it!'

I was not the only one with lips burning to make a rather cheeky remark, but Frau Doctor intervened and said to my wife: 'Frau Engel, I would be very pleased if you would come to us in Dornach—I need you as a stage eurythmist. Ilona has fallen ill and I need someone to replace her!'

We were surprised at that, for my wife had only started her training a few weeks earlier and was still only a beginner in this art. But Frau Doctor was already turning to Dr Steiner and asking him: 'Where can we put up Frau Engel?'

I report these small incidents, which are certainly unimportant in themselves, because I think they illuminate the human side of Rudolf Steiner.

We were standing in front of the mansion one late afternoon after the interviews, waiting for the car to take us to Breslau. Count Keyserlingk said to Rudolf Steiner: 'Herr Doctor, Dr Engel prescribed a remedy for me—I often do not get enough time for sleep—you are said to have mentioned a remedy which has the effect of making a short nap give the requisite refreshment.' Rudolf Steiner turned to me and asked: 'In what potency did you prescribe Uzara?' I supplied the necessary information and Rudolf Steiner explained to the Count how it should be taken.

A few months earlier the Count had come to me because of his ever-recurring frontal sinusitis. At that time I had just read that one could simply place a patient suffering from bronchitis under two electric sun-ray lamps in order to 'dry them out' and I thought that one might try a similar treatment to the frontal sinus. I told the count where he could buy a lamp of that sort and offered to write down the name of the apparatus for him. That would not be necessary, he said. Yet he nevertheless forgot what such a ray-lamp was called and the salesman convinced him that it had to be a 'Sollux' lamp. It was bought and the treatment led to complete success. Rudolf Steiner wished to be shown the lamp straight away and said he thought that Frau Doctor's larynx could be treated by it.

The car had meanwhile arrived and we got in. Rudolf Steiner sat next to the Countess, then came Wachsmuth and I. We drove in silence. That is not right, I thought and in order to start a conversation I said: 'Herr Doctor, I have heard it said that visitors to Dornach first have to pass through a "kamaloca"—is that so?' Rudolf Steiner looked surprised: 'You were only there last Easter were you not—was that what you experienced?' 'Not in the least, Herr Doctor, I felt very well.' 'Well,' he said, 'it has changed a lot since the Christmas Gathering, don't you think so, too, Dr Wachsmuth?' 'I always liked it very much,' came the answer. I recount this because, in the case of such a trivial matter as my silly question, Rudolf Steiner pointed to the significance of the Christmas Foundation Meeting. Then he turned to me: 'Why are you actually here?' 'Where else should I be?' was my embarrassed answer. 'At Frau Doctor's Speech course.' 'Herr Doctor,' I replied, 'you know that until half an hour ago we were both still interviewing patients.' 'It is true, you should be at the Speech course!' he persisted; and in some strange way I felt blessed by him: Rudolf Steiner reprimands me—he does not spare me, he can express himself quite openly to me. That was a great honour.

To conclude the conference we had a social gathering in the 'Mathiaskunst' building in Breslau. There were artistic items, speeches, singing and celebrations. But Rudolf Steiner was visibly strained by the turmoil. His hair fell across his brow, upon which pearls of sweat could be seen.

Just previous to that I had been having a consultation with a lady patient, who had had an eye removed some years previously. Now some unpleasant symptoms were showing on the other eye. I considered the condition so serious that I told her I would try to arrange a short consultation with Dr Steiner during the festivities. I managed to do this, and Rudolf Steiner gave her some immediate advice. Not content with that, however, the lady now pulled out a sketch book: 'Here are some paintings by a medium in a trance and Herr Doctor should kindly look at them.' With obvious reluctance Rudolf Steiner turned over a few pages—he had no comment to make and appeared to be completely exhausted. Then an anxious father came and addressed himself to me. His little son had cystitis and a high temperature—what could Dr Steiner advise? With that he turned directly to Rudolf Steiner and I heard Rudolf Steiner say: 'Apply hot poultices with...' For a moment the exhausted Doctor could not find the word—'...with linseed'.

I felt deeply ashamed that we had exploited Rudolf Steiner to the absolute limit.

The hour of departure had arrived. A group of friends had gathered on the platform in Breslau. A traveller looked with amazement at the scene and asked: 'Who is it who is being celebrated?' 'It is Dr Steiner' someone said. 'Aha, the famous rejuvenation researcher!' (He must have meant Professor Steinach.)

The train started to move. Everyone waved and called out to Rudolf Steiner. Karin Ruths, a young girl [see next account but one], ran beside the compartment to the end of the platform, waving and shouting goodbye until the train disappeared from sight.

Wilhelm Rath

In an address which Rudolf Steiner gave during the Agricultural Course in which he paid tribute to the hospitality of the Keyserlingk household, he also mentioned those who 'blew in'. I was one of those. I can certainly say, however, that once the 'storm had abated', a useful ground had been prepared in which the seed which this course had sown in me sprouted and eventually bore fruit. It still took seven years till my decision to become a farmer ripened.

When I stood in front of Count Keyserlingk's house in Silesia on 7 June 1924, I had no idea that a special course was about to begin. I only went there to deliver a letter to Rudolf Steiner. That came about in the following way: I had moved house from Berlin to Stuttgart during Easter 1924 in order to take over the secretaryship of the 'Free Anthroposophical Society'. This had been founded with the encouragement of Rudolf Steiner in February 1923 to be the bearer of an anthroposophical youth movement. It was one of my duties to visit the youth groups in Germany; and so, as a matter of course, I visited Breslau where, in addition to a course of lectures for members at Whitsuntide, a talk for young people had been planned. This was then extended to three talks.

Shortly before my departure I had been a guest of Walter Johannes Stein in Stuttgart and told him of my intended journey. He had handed me a large envelope containing a letter as thick as a book and closed with many seals. I was to hand this package to Rudolf Steiner. It contained reports of spiritual experiences which Dr Stein did not want to entrust to the post, and he impressed upon me that I should always carry it with me and not leave it in my suitcase, nor in the railway compartment, nor in my lodgings and should only give it into the hands of Rudolf Steiner himself.

The letter would not fit in my breast pocket and so I was rather burdened with it. When I arrived late in the evening in

Breslau, I straightway asked where Dr Steiner was living. I was told that it was at Koberwitz with Count Keyserlingk. I decided to go there next day by the first train.

When I arrived at the little station it struck me that a lot of people were alighting there. I followed them to the beautiful house standing in a great park. When all had gone into the house Dr Wachsmuth was still standing by the front door. I asked him if this was where Dr Steiner was staying. He confirmed it, but added: 'But what is about to take place here is only for farmers!'

At that moment I noticed Rudolf Steiner coming down the stairs. I went up to him and handed him the fat letter with greetings from Dr Stein. Herr Doctor took it in both hands as if weighing it and gave me a look as though he was expecting me to say something. So I plucked up the courage to say that as I had got here, and although Dr Wachsmuth had told me that the lecture was only for farmers, I would like to ask him if I could attend without being a farmer. Rudolf Steiner's eyes lit up and he said merrily: 'Well, if you're not a *Landwirt* [farmer], at least you may be a *Landstreicher* [vagrant]. (And that was a fairly accurate description of me during the next few years!) Then he put his hand on my shoulder, and guided me into the room like that past Dr Wachsmuth, saying to the latter: 'We will give him a ticket, don't you think?' So I sat down among farmers, received my ticket and was able to attend all the lectures of the Agricultural Course.

That was of decisive importance for all my subsequent life. I had not brought any questions with me about practical agricultural methods. Agriculture was something very remote to me at that time, for I had been studying German Philology in preparation for a profession as librarian. I had given this up for the time being when the Youth Course took place in 1923, so that I could work for the anthroposophical youth movement. That might be why the spiritual basis of the present course struck me immediately, in a very stirring way.

The cosmic source of energy for plant growth; the hidden spirituality of limestone, clay and silica; the components of albumen, described as the bearer of creative activity—all that appeared to me like the revival of ancient mysteries—like a modern way of initiation into the eternal mysteries of great, creative nature. A quite new picture of agricultural work was formed in me, a picture which could not only become a way of knowledge of the spirit at work in nature, but also a priestly activity of redemption for the creatures of the earth. For this reason the esoteric 'Address to Young People' with which Rudolf Steiner concluded the course for us young folk, became an experience which later led me into the farming profession. Many participants may have had a similar experience. For the words of Rudolf Steiner, issuing from spiritual depths, showed us that even in farm work the spirit may be sought. Anthroposophy shows us the way, because through it one no longer strives only for intellectual understanding. 'We must learn to understand in our hearts that which converts the spirit which is as yet only "thought", which is alien to nature, into the spirit which has been worked upon and which now leads out again to the realities of the natural world.'

The spirit which has been worked upon, however, allows us to see the activity of the farmer—the spreading of manure, ploughing, sowing and reaping, even the baking of bread—in a new light. Just as the farmer helps to unite heavenly forces with those of earth, so do those who strive for knowledge do the same. The farmer manures the earth so that it can become receptive to the working of cosmic forces which want to be active in the growth of plants and the ripening of fruit. The seeker of the spirit 'manures' the fields of the spirit by changing anthroposophical thoughts into devotion, so that divine powers may continue to illuminate and fructify man's spirit. The meditating human being thus transforms shadowy abstract thinking within himself into living, colourful, pictorial thinking. This thinking becomes a conversation with the creative forces! Digging, manuring and sowing become a symbol for inner activity, the care of the earth is a symbol for the care of the soul, and the patient waiting of the farmer for the sprouting and ripening of crops becomes a symbol for the mood of expectancy which is a prerequisite for the experience of all higher knowledge. This conversation of Rudolf Steiner with the young people, to which he gave the character of an esoteric class lesson, because he recognised the holy seriousness of our hearts, became for us the knowledge that there is also a path to the spirit through work, through practical work!

Agriculture, enlivened by the Koberwitz Course, does not only have the possibility of retaining the fertility of the soil—and by way of the food thus grown, also the creativity of man's thinking—but it also bestows on the farmer a deep satisfaction in his work. For through this kind of work on the land, cultural life can develop! When young people engage in it in this way with vigour and efficiency, it surely leads to an experience of the spirit. Young folk may not set much store by study in the old sense, but this is precisely why they can become the bearers of social culture.

In my case these experiences gave me a new understanding of the ancient mystery currents which, taking their start from Zarathustra, by way of the Demeter Mysteries of Greece, the Platonic vision of the 'Goddess Natura' in the School of Chartres and further through Rosicrucianism, finally led to this new anthroposophical agriculture.

I not only became conscious of the fact that work on the land had become a new career for striving young people, but that work on the land in this sense would have special social tasks to fulfil for Central Europe in times to come. During a tour of the estate a young member of the Keyserlingk household told me about a conversation between Count Carl Keyserlingk and Rudolf Steiner. My memory of this conversation was that Rudolf Steiner foresaw for Germany a future of 'falling chimneys, cities lying under rubble and ruined factories and that a plan would exist for the whole of Central Europe to be turned over to agricultural land'. But under such circumstances there would be a need for establishing 'agricultural-based centres for nurturing the spirit', where young people could be brought up in monastic-like settlements. And I recollect having heard that these settlements would be like 'oases in the desert'.

When I heard that, I said to myself: that is also one of the tasks for the anthroposophical youth movement, for we young people will have to live through that future. Again and again during the following years I have discussed with friends how this could come about until—as the signs of decline became ever more evident—I took the decision to become a farmer myself. My wife agreed with this decision, as she had also been connected with the anthroposophical youth impulse ever since the Youth Course had been given. A share in an inherited legacy which she had received made it possible for us—after five years of practical training in agriculture had been completed—to acquire a large farm in Kärnten, Austria. This happened on the same day that the Anthroposophical Society was banned in Germany. There we had the good fortune to survive while Central Europe fell into the abyss. On this farm, which grew—if only to a modest degree—into a centre of care for the spirit, the fruits of the Agricultural Course can continue to ripen.

Karin Ruths-Hoffman

When I think about the Agricultural Course, and at the same time about my youth, I can see Dr Steiner as he spoke to us and turned his attention to us. His expectations were great in comparison to what we fulfilled of them—yet, if you think about all that he had to offer us, he was surely reaching out to more generations than just the one sitting before him.

I had been sent to Count Keyserlingk as the representative of a group of young teachers, with the request to be allowed to take part in the Agricultural Course. 'Tomorrow, when Dr Steiner arrives at the station, we will ask him,' said the Count. So I was presented to Dr Steiner at the station in Breslau for the second time and he gave—as he had done before when I had asked to be admitted to the Waldorf School—a very friendly 'yes'.

And then came those very beautiful Witsuntide days. My father was there too. He had got permission to attend from Her Excellency Frau von Moltke when he met her at the Hotel in Breslau. He was a native of the same town in Sweden as she was.

So we all sat in the early mornings in the same Koberwitz train: important estate owners, eager gardeners, enthusiastic or critical young people. All thought of themselves as 'Koberwitzers' and chatted together in a free and easy way. Then we listened to the morning lectures, heard about 'heavenly animals', the pigs, so that the vegetarians' hair stood on end. One had to radically revise one's ideas. During the break everyone engaged in animated discussions and Dr Steiner walked around very happily among us.

Of the young people there it was Rath and Grunelius who took the initiative to ask Rudolf Steiner for what later became known as the 'Address to Young People', which was preceded by the two addresses in Breslau.

Rudolf Steiner began by stressing his own youthfulness

and saying that he did not in any way wish to be treated as a 'father'. We asked him in what way we should work, for we had already had quite a lot of arguments and even a real scuffle on one occasion. He answered that the method to be employed is not so important. What is important is that one or two enthusiasts are present. This would lead others to join in. If one is to introduce people to anthroposophy, one has to have enthusiasm. If he himself had had a profession to which he had been tied, anthroposophy would never have come into existence. He enjoined us strongly never to play truant from our meetings, but to stick together with an iron determination.

How happy we were to hear from Rudolf Steiner that he would welcome it if anthroposophy and the youth movement could come together, for they were akin to one another. The youth movement with its experience of community arose through the fact that on our way down to earth we had all experienced Michael. 'You had a vision of Him in golden armour, holding a sword, with pointing hand and in his spiritual armour.' The way he said this stirred me very deeply. There were those among us who had never heard anything like that before!

The second address began with a question from me about experiencing nature: if one walks with others in nature one experiences an intensification and enhancement of oneself. But when alone in nature, one can experience it as a voice or a piece of writing which one wants to understand. What is this voice saying?

In his reply Dr Steiner spoke about the birth of tragedy out of the spirit of music, as described by Nietszche. But it was only from his third Koberwitz address that I had my answer—through what he said about nature: about the gods having been enchanted into the rocks and stones, about our brain, the 'manure heap', about understanding the language of the birds, Siegfried-remembering and being able to 're-dream'.

In his first address Dr Steiner said he was glad that nobody was taking notes—at which, in shock, the pencil fell from the hand of one person who had been writing.

Now these addresses have been printed, but very much abbreviated. A lot less is contained in them than what I still remember. But what concerned Dr Steiner was apparently that the tone, the warmth and dynamic should be just as strongly absorbed as the written content. He even said that someone had once reproached him by suggesting that the book *How to Attain Knowledge of Higher Worlds* was a product of the typewriter. In actual fact, he said, he had 'walked' it—that is, had conceived it while tramping the countryside.

These hours together left a deep impression on each one of us. We had been formed, confirmed, transformed; and that was immediately apparent. I remember one young man who suddenly stood up and started to sing about a dead comrade who had belonged to the German *Wandervogel* movement. Rudolf Steiner took this personal eulogy quite seriously and said to him that we could all only fulfil our tasks if we took those who had died young into our consciousness. They were eager to help us in our aspirations with their youthful, unused etheric forces. What conclusions can be drawn from such a suggestion!

The intensity of these hours, together with the Karma lectures in the evenings, must have had a tremendous spiritual effect.

And then there was something else of importance which one cannot read anywhere—the strictness with which Dr Steiner treated us when it proved necessary. On one occasion Count Keyserlingk had to stand up at the start of an evening event and call out the names of those who were required to come forward to relinquish their membership cards, because they had behaved in an undisciplined and disturbing fashion. This branding process made a very strong impression on us.

The third, 'the Koberwitz Address', was a very serious affair. To begin with we had to wait for half an hour in the cold park for someone who had overslept, and we were moved by the fact that Rudolf Steiner, who could be so strict, was so considerate about this. This gave us an example of how to conduct ourselves in everyday life—never in a narrow, stereotyped fashion.

When Dr Steiner shook each one by the hand at the end, all was drawn to a conclusion, so that we knew that this had been an esoteric gathering, a spiritual event—as though a fire had burned among us, a fire of sacrifice.

It often appears uncanny to me today that the memories of those who were present at the course can be so different. What remains, though, is not objective facts but the subjective relationship which the individual has formed with what took place. This has worked on in each person—and is still doing so. This is true even of those who were not present but have nevertheless participated by reading about it.

A prelude to this Whitsuntide Course were the preparations for the founding of the Christian Community, which was undertaken in Koberwitz by Geyer, Rittelmeyer, Bock and other priests, who had been frequent and long-term visitors there. This was an intimate prelude, whilst that other, larger-scale one—the Threefold Commonwealth Movement—had almost ebbed away.

Bock was often impatient at that time because things progressed so slowly. He glowed with ardour to get started, as did also the young farmers, and could hardly bear to watch how the older folk proceeded at an outwardly slower pace. And yet it is so important, as we see today, for all who have embarked on spiritual tasks to go forward in unison.

In this connection I can think of a scene which took place in Stuttgart before the Delegates' Conference in early 1923. There are very few today who remember what Dr Steiner told us on that occasion: 'The Anthroposophical Society has behaved towards the Christian Community like a father to whom a child has been born during his absence, and who then returns home, first has breakfast, reads the paper and goes for a stroll before enquiring after the child.'

Rudolf Steiner said this long after the well-known lecture which so many people took as an occasion to emphasise the differences between anthroposophy and the Christian Community. To the above metaphor of the absent father he added a second one: a married couple had their child christened Johannes, according to genuine anthroposophical principles—but then they forgot it on the railway platform! With that Rudolf Steiner turned to Frau Doctor, who laughed until the tears rolled down her cheeks.

And still another situation springs to mind in this connection. A friend asked Dr Steiner how one should combine being an academic and a young anthroposophist. Dr Steiner replied: 'You will have to write interesting doctorial theses—you will have to become a genius at taking interest'—he meant by that an interest which stops at nothing and extends to points of view which contradict one's own view. One cannot embark in every boat, but one can be interested in everything.

And so we can make interest in each other into a spiritual bond which unites us all across the generations.

Dr Steiner once said to us: The threefolding of society did not fail because it was a wrong idea, but because its adherents failed to understand it—and I hope that from your ranks, from Waldorf pupils, will come those who will develop a proper understanding.

He thus pointed to a future which has still not arrived. Our individualism is still not strong enough to develop 'a genius for taking interest'. We still think in terms of quantity rather than process, although the latest discoveries of science lead out of space into time, into process. Our thinking hobbles on behind. We have now arrived at a transition from a Mars to a Mercury Age, which expresses itself in battles leading beyond the individual to brotherhood—to tolerance, flexibility and interest in others.

Through the way that it reunited life-sustaining nature with the spirituality which underlies all things, the Koberwitz course was able to show, to each one of us who had the good fortune to be present, perspectives extending far beyond the bounds of agriculture. These are the tasks which confront young people today, as they did then, as a counterbalance to the false path of pure materialism—as a challenge and as a future goal.

Helmut Woitinas

As a young assistant gardener in search of my spiritual home and the meaning of life, I was attracted to the *Wandervogel* movement.

There were several branches of this movement: the Goths, the Teutons, the Good Templars, the Warrior Templars—all, however, wanted to experience nature and get away from philistinism.

When the first anthroposophical lectures in Breslau were given in the autumn of 1919, a youth group formed of artisans, students, pupils and employees met in the evenings. We read, had discussions, went for walks and acted in the Oberufer plays at Christmas and in the Redentin Easter play.

One Sunday, while we were out walking, three young people joined our party. It was a hot day and we went for a bathe—then it happened that one of the three was drowned before our very eyes. On the other bank of the river was a party of Warrior Templars who were witnesses to the accident. Some of them swam across the river and took part in our desperate search. But we were unable to find him and had to undertake the painful task of informing his mother. The Warrior Templars accompanied us.

Through this shared experience we often met together—they brought others along with them and our group grew larger. We found our way to serious anthroposophical work.

Then, when Rudolf Steiner came to Breslau, we had the unforgettable experience of his support and spiritual direction in our search for nature, our protest against narrow-minded bourgeois ways of thinking, and our desire to pursue good and valid aims.

The 'Free Anthroposophical Society' came into existence through the meeting together of young anthroposophists; and as, under Rudolf Steiner's guidance, the kernel of truth emerged in all branches of study, we saw before us the ideal form of humanity for which we had been longing. Goodness of heart and perceptive understanding streamed from him. The world fell into place and a spiritual background was revealed, much more wonderful than anything we could have imagined. Each one of us received impulses which enabled us to proceed further in life.

Then I was allowed to take part in the Agricultural Course and that brought me together with people I would otherwise hardly have met at such a time of extreme contradictions—strikes, unemployment, inflation and political arguments—which were then convulsing Germany. There was a sudden universal interest in telepathy, hypnotism and Spiritualism. Anthroposophy was made public but its answer to the social question, the Threefold Social Order, was not accepted because there were insufficient courageous politicians to promote it.

Fifty years have since passed. Germany has been engulfed in the Second World War and has since invested all its forces into economic reconstruction. It has adopted American standards; and if one asks what young people are like today, it becomes clear that whereas the problems occupying our minds then were much more in the realm of feeling, personal matters, questions of vocation and the life of nature, today the experience of the young springs mainly from the will. Intellect and will have become much stronger, and there is an urge, expressed either in endless discussions or in brutal riots and tumult, to overthrow the present order. Every such movement spreads beyond all bounds. Not so much personal problems, but questions of changing consciousness, of living together with others, press upon us far more fiercely and insistently than before. This intellect and will are more intense, but also more vulnerable than before, because they are no longer under control. They are becoming independent! The striving to understand life is more desperate, and ways of dealing with situations have become more radical. And yet the feeling of responsibility towards humanity is taken far more seriously.

Our overconsumption has not only repelled young people, but also weakened them, so that they are largely unable to carry out their programmes against all the odds, or think things through in a vigorous, living way.

The middle realm has become very dispersed and weakened by cold intellect on the one hand and unrestrained will on the other. Young people feel alienated by soulless materialism, and yet dependent on it, and take to the road to look for humanity and warmth in others. There they meet intoxication; or ideology in the form of glittering objectives, which in reality are set upon destroying the middle way, the ego of man. Many lack this middle realm—which enabled us at that time to recognise anthroposophy and one other. It was therefore easier for us, for we had something we could trust.

Rudolf Steiner called upon us to forge inner weapons! But is seems to me that the implication of his ideas was not fully understood—even then—for the social impulse implicit in the Agricultural Course would suffice to solve the burning questions of our time. At that time, however, some of the farmers had only just become aware of these problems, brought to their consciousness by the trend made clear by the development of chemical fertilisers: land treated in this way needed more and more of these fertilisers, the fertility of livestock decreased and likewise the quality of foodstuffs. Pests increased to a horrifying extent and had to be destroyed with poison, which endangered birds, the water supply, and ultimately the health of mankind. It could have been foreseen that the increase in crop volume would be accompanied by a decline in quality. Today a natural appreciation of quality has already almost completely disappeared; any producer is able to sell inferior goods by means of empty advertising slogans!

This development, which was forseeable, has occurred with terrible consequence; and a call for bio-dynamically grown products becomes ever more insistent. Yet a knowledge of the harmony within nature and the spiritual background from which it springs—that is to say an ability to handle living processes, or, more succinctly, a dynamic sense—is only present in anthroposophy. And it is only on this basis that the burning questions of the day can be solved! Intelligence and will are both necessary, but without spiritual commitment—without the middle realm of the heart in which what is personal and what is objective are united and in which courage and constancy are able to grow—we shall not be able to progress.

At that time, however, what was subjective and what was objective were merged in a great common experience for us young people.

I experienced the especial favour of often being invited by Countess Kaiserlingk to sit at the dinner table along with Rudolf Steiner and other guests. One can imagine how it felt for a young gardener's lad to sit next to the lady of the house—and if I failed to help myself lavishly to what was provided, to have her replenish my plate with the tastiest morsels.

Every day was a new experience. Rudolf Steiner showed us the spiritual background underlying the great living organism of the earth, and from that he developed the practical means to work with it. The harmony of life and death, expansion and contraction, of the seasons, and the mutual stimulation of all earthly processes was unrolled before us like an immense panorama. I, who had been taught that weeds had to be eradicated, now learned how useful some of them were. We learned, for instance, that Chamomile grows where lime is lacking—not because it contains lime within itself, but because it is able to stimulate the lime-building process within the soil; or the horsetail [equisetum], itself dry, almost brittle, grows where water accumulates in order to regulate the harmony between what is too wet and what is too dry. Stinging nettle regulates the harmony in the iron-process. Rudolf Steiner, with his insight into the spiritual background of nature, gave exact instructions about how, for instance, one could enhance the forces and properties of weeds in conjunction with certain animal organs. By so doing the farmer introduces them into the earth and cosmos in such a way that active, radiating ferments are created which can be used as sprays and additives to manure and compost, so as to enliven the soil.

It struck me as a special miracle that quartz, which according to modern science is the most sterile of substances, can be transformed into a light-giving process by association with a cow's horn, and is then able to extend its healing influence into the plant (e.g. to counteract the rotting process).

These methods, developed out of spiritual insight, but of immediate practical relevance, could be an antidote to the disorder and disease which man has introduced into nature in the 20th century through his technology. Man will then be integrated and involved in nature once more, and will stand as a mediator between the earth and the universe. For he is not the mere product of his surroundings, but the one who creates his environment! That was a thought which, at that time of chaos, showed me that life is worth living, and gave me an understanding of much which I had formerly dismissed as a 'pretty tale':

> How heavenly forces which ascend, descend
> Their golden vessels one to another lend.

Those are not just pretty words—they are reality.

Erna van Deventer

The healing cure for our earth, which sustains us but which has become sick, and the seed from which living thoughts can grow to provide us with the practical means of effecting this healing—was given us in Koberwitz in Silesia through the Agricultural Course. And the name Koberwitz calls to mind thoughts associated with the earth, with something homely: a lovingly arranged household with its garden, park and fields; the host and hostess, Count and Countess Keyserlingk; and Rudolf Steiner! It was as though goodwill shone down upon us.

I came there as a eurythmist and knew about work on the land only as much as Rudolf Steiner had told us in Dornach, in lectures, advice and meditations, about the nature of plants and the tasks of medicine.

I will have to recount something about myself in order to describe how deeply I was impressed by the wisdom of Rudolf Steiner.

He was so aware of what lay in store for the young people that flocked to him that he gave to my future husband, a young anthroposophist facing his doctor's examination, and to me, a eurythmist, common tasks which shaped the whole future course of our lives.

Through the fact that my destiny brought me into contact with Rudolf Steiner during my childhood, and that it was he who called me to eurythmy and, at an early age, gave me the possibility of experiencing what healing meant, I was led to H.A.R. van Deventer, my life's partner.

I was able to speak with him about things which would otherwise have remained closed within myself; from childhood on I had regarded trees, grasses, mushrooms and all plants as though they were mirrored aspects of something within the human being. Thus it was of great importance to us that we were able to put questions to Rudolf Steiner whilst the first bio-dynamic experiments were being carried out in Dornach, long before the course was held. And we had the wonderful experience of hearing him give answers, which exceeded our expectations, to tentative questions such as whether one could develop therapeutic eurythmy out of artistic eurythmy—curative eurythmy! I mention this to show how seriously Rudolf Steiner spoke to people of our young, inexperienced generation.

According to Rudolf Steiner's wish my husband who had been at the course for young doctors in 1924, was allowed to tell me all that had been discussed there, including the meditations, although I was not a doctor myself. The 'will to be a healer' was an expression which fascinated us. This encouraged us to ask Rudolf Steiner how we could find the way to the healing spirits which worked within and around the plants.

Rudolf Steiner then arranged a consultation with us and asked us how we envisaged this healing process. But we did not know ourselves—we had come to him in order to find out! Then he smiled and was silent for a while and we thought he would dismiss us like children who ask silly questions. But it turned out differently: he asked us how we had come to the idea of healing and we said that through a difficult destiny we had come to a time of great loneliness, and had had an invisible helper whom we called our 'Faithful Eckhart'. Rudolf Steiner did not laugh at us, but said: 'This Faithful Eckhart is a stimulus towards your work together' (of doctor and curative eurythmist). He explained that it was through blows of fate, such as we had experienced, that people were made aware of things which they would not otherwise know about. We should test the advice given us by our Faithful Eckhart against that which we heard and learned through the lectures. My predisposition to experience people as plants, and plants as the reflection of human qualities of soul was a good preparation for 'healing'.

After the Christmas Foundation Meeting we went to Holland and Rudolf Steiner gave us the following task to help us with our search for the nature of the healing spirits:

'You are both to go every morning, at the same time, to the same place, where there are meadows, trees and water. Observe a certain place—always the same one—under all conditions of weather. Make a change of location at the most only between summer and winter. Study the trees there and the shrubs and grasses, the forms of the clouds, the shadows and the rays of the sun on the leaves every morning. After your observation do your meditation about the 'healing spirits'. And what you then have to try to do is to look at the plants, not by abstract study, but with living observation. The shapes of the roots, stalks and blossoms must speak to you—the plant must reveal its inner nature to you! Include in this the movement of the leaves, of the stalks and even of the roots. Connect that with the eurythmical movements—they are the same as the gestures and the movements of the plants. Take a fir-tree ('TANNE'): what does its gesture say to you? It really makes a protective gesture round the person standing beneath it: a T, an A, an N, when it is swayed by the wind. The fact that other letters are used in other languages only tells us that a Frenchman, for instance—who calls it *Sapin*—recognises another aspect of this tree.

Or the foxglove, what does it do? Its fine, loose roots are able to sink deep into the earth as a result of the rain. Then it develops a very tall stalk which rises quite beyond the earth's gravity. And at last come the bell-like blossoms which again bend towards the earth: can one not see imaginatively how the foxglove expresses a state of equilibrium—first uniting in its roots with the force of gravity, then striving upwards like an erect candle and, in spite of bearing the weight of the blossoms, reaching out towards the light; but turning again towards the earth with its blossoms. Can one not see by looking at it that it must have a harmonising effect on the heart, alternately uniting itself with gravity and then with the sun, and keeping itself in harmonious balance? In this way meditation must become a reality! Sulphur and salt, with mercury in between, work and become a reality. In this way one unites oneself with the basic elements of the plant's existence!'

We travelled to Holland with these instructions and began our plant studies in March 1924. At 7 o'clock every morning we cycled in the meadows beside the Rhine—at 8 o'clock my husband had to be at the clinic.

Perhaps young people of today are better able to appreciate how helpful this study was. It was a dramatic way to study—every morning different weather: storm, snow or spring sunshine. No sooner had we grasped the changes which rain brought about in the plants than a thick mist shrouded everything the following day. We had to connect these changes with the mood of the weekly verses from the *Calendar of the Soul*: 'When from wide world-spaces'; 'When from inner soul depths'—yes, in this way our thinking became pictorial. And when I did eurythmy to the sounds of 'world-widths' (Weltenweiten) and 'soul-depths' (Seelentiefen), then these words became deeds, became visible gestures of the gods on earth! They became the key to understanding forms and movements in the mineral, plant and animal kingdoms. 'Thereby you will be able to decipher what is beneficial or harmful to man.'

One is reminded in this connection of the Vienna lectures given in 1914: *Occult Reading and Occult Hearing*. Which of us was advanced enough to recognise in eurythmy the visible speech of the cosmos in the realms of nature, and to see its true reflection in human speech? Sometimes we had a dawning recognition of St John's Gospel—that it is the WORD by which everything was created! And when we do words of cultic significance in eurythmy, when we are told that the word EVOE for the Greeks still meant: 'We seek one another and have found one another'; and when we are told exactly how we should perform this exclamation, then the way to the consecration of speech is shown us. (And the path of healing by means of speech—curative eurythmy—was first indicated in 1921.) Is not the genius of language right to affiliate the word 'holy' with 'healing'?

Let us return to the tasks given us by Rudolf Steiner in Dornach in Autumn 1922, when the first experiments were being made with cows' horns. Trained farmers could of course understand Rudolf Steiner's explanations of these experiments better than we—I and a few other eurythmists. Yet stirring with a stick in an old bucket—until our arms began to ache—was something for us! The biological aspect was not clear to us, but the dynamic part was—for one of the meanings of the word eurythmy is 'good rhythm'. Our eurythmy movements were a submission of our own wills to that of another, a cosmic will, which is able to unite itself with us to our benefit—just as the rhythmic stirring of the preparations serves the re-enlivening of the earth to which one entrusts the seeds.

Perhaps it takes someone connected with eurythmy to fully experience the subtle effects of such movements of the hand, connecting it with matter. Whoever has taken part—as in Munich in 1913—in the movement of the sylphs in the Devachan scene of the *Mystery Dramas*—raising and lowering one's arms with veils for seven minutes on end, dipping them down into gravity and raising them up again high above gravity—would feel at home with the stirring motion involved in making the preparations.

And during evening question-times in the office of the First Goetheanum, when discussion turned to the lemniscate movement in connection with astronomy, Frau Baumann-Dollfuß and I received a flash of inspiration. It was this particular movement which gave us occasion to ask Rudolf Steiner about therapeutic eurythmy: how could macrocosmic laws become a means of healing human beings?

Does this not all go to show that there are two possible ways of understanding: an intellectual and a dynamic way—the scientific way and the way of eurythmy? And how beneficial has the knowledge of matter since proved which resulted from the latter approach. We have had manifold confirmation of the healing effect of the ideas developed in this way, especially in education and therapy! It was a new experience we had by following Rudolf Steiner's advice; not just to study the plants botanically (which went without saying) but to study them in inner experience: to understand them by experiencing them! 'Do not forget that every plant has a front and back, an above and a below! The climbing plants strive heavenwards, the ear of corn bends down towards the earth! Make a drawing of that. Perform the movement of burgeoning and withering in eurythmy! Those are worthwhile exercises, for the sounds are what clothe the words—e.g. *Hinauf* (upwards), *Hinunter* (downwards), *Leuchtkraft* (shining lightness), *Schweremacht* (heaviness)—let the words produce a picture in you!' That is what Rudolf Steiner told us.

To conclude there is just one small example of how Rudolf Steiner taught us to grasp the possibilities of healing through the experience of the healing spirits: we should study the horse chestnut tree, its prolific foliage and blossoms in which one can almost see the etheric forces. We tried to portray the leaves through eurythmy. To do that one must be familiar with the rich store of eurythmy movements and should not hang onto a fixed scheme. Wherever there is a crossing-point there is an E (eh). What a lovely 'Eh' sounds when the outflowing breath and the slightly parted lips cross one another! Thanks to such subtle concepts it occurred to us that the horse chestnut in spring must be a remedy against faintness, overwrought nerves and cramps of all kinds. We told this to Rudolf Steiner and Frau Dr Wegman—and how universal has this therapy now become!

How grateful we were to our destiny which had brought us such tasks as these, how grateful to our 'Faithful Eckhart' and above all to Rudolf Steiner who had taken us seriously and was our teacher. And how intensely we, who had set out on this path, experienced Koberwitz, where a flaming torch had been handed to us, which we now hand on to the younger generation.

That is why this contribution has been written.

While writing this the authoress had a battered old notebook beside her in which Dr van Deventer had recorded all his conversations with Rudolf Steiner.

Count Adalbert von Keyserlingk

My mother took me to see Rudolf Steiner in the turbulent times after the First World War when she did not know what to do with me any more.

She had got to know him a short time previously and had often spoken about him at home and about his ability to see things which people ordinarily do not see. So I was full of expectation, but also a little afraid that he would be able to look into the innermost secrets of my soul. We therefore went to Motz Street and were shown into a small room with a white tiled stove. Rudolf Steiner came in a few minutes later and gave us a friendly greeting and then I was sent out of the room.

My mother had made out a whole list of my difficulties and characteristics and wanted to read them out. As I had no alternative but to wait in the little corridor outside I heard every word. Then Rudolf Steiner came out again and led me into a small dark room where there were bookshelves with documents. I had to wait there for some time.

It was at a much later time that my mother told me what she had said about me: that I did not want to learn, could not spell properly at the age of 14! But Rudolf Steiner had laughed and said: 'But that is very healthy!' And when she interposed to say that I would not be able to pass my matriculation exam, he answered 'By then I hope that we shall not need matriculation any more!'

She told him further that I had been interested in nature study since early childhood and in refined thought—but that I could not remember poems or stories or learn vocabularies by heart: 'He is very lazy as soon as he has to think, he is unable to apply Latin rules and gives up completely on foreign languages!' But Rudolf Steiner pointed out to her that bad memory showed a disposition for spirituality.

Then my mother told him that she had always seen me

surrounded by death from early childhood, but that now it was better. Still, I did not yet have the resilience of youngsters my age. Half an hour of gardening or studying would suffice to make me turn pale. So how would I be able to get through school life?

Rudolf Steiner thought it would be good to let me have private tuition for another two years and gave exact instructions regarding my health.

When he came to fetch me again he put his arm around my shoulders and stroked my head. Then I had to sit by the white tiled stove and he asked me how I liked school. I hummed and hawed because I did not go to school at all. I had been brought up entirely in the country and had played with gnomes as a small child. Then I had become accustomed to work on the land with two horses, Max and Moritz, had carted corn and turnips, harrowed and ploughed, also with oxen, and had ridden the horses to water. I had worked in the woods with the forester and with my dog and helped to clear the ground for cultivation, marking the trees for felling and had observed many things on my own. I had not learned anything—at least I had no intellectual knowledge, for the Pastor belonging to the estate and the private teacher from Breslau were only too pleased to consent to my mother's wish to leave me in peace.

When the Waldorf School was founded my mother wanted to enrol me there, but Rudolf Steiner advised her to first send me for a year to Haubinda, to the intermediate grade of the Lietsche Boarding School.

What was I to make of Haubinda? It was there that my mother became acquainted in 1920 with the teacher Christoph Boy. I boarded with him, and after only a few months Herr Boy wrote to Rudolf Steiner. He received a telegram in reply appointing him as a Waldorf teacher—without any probationary period—so that he arrived there long before I did. He brought a colleague, Herr Kilian, with him and even a third teacher—Parson Seusing—was to have come to Stuttgart, but for domestic reasons, he never came. Thus my short stay in Haubinda had fulfilled a definite karmic purpose.

In Stuttgart, too, I lived in Christoph Boy's house, until the end of my schooldays.

That first half hour in Berlin is something I shall remember for the rest of my life: a consciously experienced inner conviction was formed. My attitude towards Rudolf Steiner changed in a very positive way, so that when we took our leave of him I felt that I belonged to him. But another change took place in me too: my attitude of soul which lived completely in union with nature grew to a kind of awakening within myself. That does not mean that I did not continue to dream on for a long time—I knew, nevertheless, that my path was one of schooling through Rudolf Steiner.

Christoph Boy wrote to me at Haubinda and said I should come there quickly for it was a pity to miss even one day at the Waldorf School. When I arrived he received me in the playground and asked me which class I preferred. I said I wanted to be in the same class as Walter Molt. According to my age I should have been in the class above—but putting me into the 9th class allowed me to stay at the school two years longer.

I now saw Rudolf Steiner many times. My heart was wholly his—as were those of all the children who clustered around him in the school playground. He also took the same route to school as I, and often met me, but he was never alone. There were always grown-ups with him in animated discussion.

When Rudolf Steiner was in the school he visited the classes and listened to what was being said. But often he stood to one side with the window at his back, took up the subject and continued the lesson himself. I clearly remember a history lesson in which Walter Johannes Stein was speaking about the Crusades and about Barbarossa and his mysterious death. Slowly Rudolf Steiner stood up and started to tell about the conditions which prevailed at that time and how the people of those days had a quite different consciousness from that of people today. A vivid and colourful picture arose in me and stayed in my mind, although I had always had such a bad memory. Then Rudolf Steiner described how Barbarossa with his army of Crusaders came to the river Saleph in Asia Minor on their way to Akkon. From his horse Barbarossa saw how difficult it was to cross the narrow ford on foot. That was too slow for him so he rode into the water and came to a deep part of the river. His armour pulled him under and because of the great difference in temperature between the Asiatic heat and the cold rushing water he lost consciousness, could not regain his horse and was drowned.

Yet it took nearly two years for the news to reach Germany. News could only travel by couriers, minstrels, tradesmen or pilgrims and the people of Germany loved their Emperor so much that they did not want to believe he was dead. The saga grew that he was not dead but had withdrawn into a mountain until the start of a new epoch in Central Europe, until new impulses were required in the world. Ravens would fly around the mountain to announce to him when the time had arrived.

After Rudolf Steiner's death Dr Stein told this very same Main Lesson story in the memorial ceremony, at which so many of the pupils, especially the young ones, were in tears.

On another occasion Rudolf Steiner entered the classroom and went straight to the teacher's desk which stood by the window. I looked over towards him—every motion he made seemed of interest to me. He looked at every pupil one after the other. I had heard from Christoph Boy that he knew every scholar—not by their names, but by their places in the class—and would bring each one of them to mind every evening. I made a practice of that when I was a teacher and I found it the surest way to remember each single child.

His gaze rested on each one of us. So it also came to my turn. Yet I felt neither fear nor anxiety, only a wakeful anticipation. It was quite clear to me that he knew more about me than I knew of myself and I felt no opposition, no resistance, for I could feel myself enveloped in complete sympathetic understanding. I felt his steady gaze, saw his dark brown eyes which shone with a questioning kindness. And again I felt that I belonged to him. It was a conscious decision: just as before, my life only had a meaning if I could remain attached to him. I can now say that it was a knowledge lacking rational explanation or deliberation, an intuition which was there and remained fixed there—without emotion.

Christoph Boy once related later that he had spoken to Rudolf Steiner about me and had been told that I would often have difficulties in life because the Baltic-Eastern etheric body inherited from my father had not yet harmonised with my Scottish astral body from my mother.

Towards Christmas 1923 I was told by Christoph Boy that Rudolf Steiner had given permission for two pupils to be present at the Christmas Gathering—Alex Vivenot and me. But he did not, I think, sufficiently explain to me what was to take place there, so I returned home instead.

It is often insignificant experiences which deeply affect one's mood at this age. It must have been in the 9th class—our classroom was then in the main building. One dull, misty morning we were aware of the fact that 'the Doctor' was there—but he had not yet been to our class. I was feeling very despondent for I had noticed on my way home that I had left my coat hanging outside the classroom. I returned for it and, just as I was putting it on, Rudolf Steiner came out of the little office which was opposite our classroom. He was followed by Dr Schwebsch. It was obvious that Dr Schwebsch had not finished speaking to him, he was talking quickly, looking up while he did so and then remained standing in the doorway. Rudolf Steiner came towards me, gave me his hand in his usual way, took a few steps, turned round and with another wave to me, went back into the office with Dr Schwebsch. It was obvious that he had only come out of the room to give me his hand—he had noticed that one of his Waldorf pupils was unhappy.

In the 10th class we were accommodated right at the back of the old barracks, access to which was attained through a long dark passage leading past the eurythmy room to the school gates.

Whilst standing one day in this dark passage with Walter Molt and another classmate, Walter—who always had some mad scheme in mind—suggested that for fun we should kiss

Lilian, with whom I had been friendly since my first day at school.

Our thoughts had otherwise always been very moral and such things actually never occurred to us. And now Lilian came. Walter received a box on the ear for his pains, Max got a shove and I—got a kiss. But because this scene was not fair play at all, Lilian reported it to Dr Röschl and the latter brought it up at the teacher's meeting! The son of the school founder and the son of a well-known anthroposophist and friend of Rudolf Steiner! The teachers were in a difficult position. They reported it to Dr Steiner. I do not know what he said to them—but Christoph Boy was instructed to give me a talking to. He told me that such things should not be done in a dark passageway, but preferably on a seat in the woods.

Lilian was not cross with me, Walter remained silent about it—everything seemed to have blown over—but Rudolf Steiner was due to go to Breslau and would certainly tell my parents about it. So I decided to tell them myself and wrote them a letter to say that I had done something which had turned out badly and had even been brought up at the teachers' meeting. So then, when my mother was travelling from Koberwitz to Breslau, she asked Rudolf Steiner what I had been up to. He told her about it and said that Walter Molt was really the instigator of the affair; then laughing said: 'After all, such a thing belongs to the enrichment of life!'

In Class 11 Alex Leroi and I had been appointed class spokesmen, and as we could not get along with our French teacher—or he with us—Alex and I were asked to speak with Rudolf Steiner to ask him if we could have a different teacher.

So we waited for him on the main steps to ask when we could have an appointment with him. He came out and quickly descended the steps. Alex ran after him and I—remained standing. For at that moment, on seeing Rudolf Steiner, it became clear to me that I was the worst pupil in French and had not even improved with extra lessons—I was too lazy for that—so how could I complain about a teacher whom I was even quite fond of?

It was quite different with Alex. He spoke German and French just as fluently as his mother tongue, which was Portugese. He was the best in the class and was quite justified in complaining about a teacher who had less command of French than he himself. But I?—Oh no!

I can see him now with Rudolf Steiner on the steps as he received an appointment. Then he returned to me—in a mighty rage! He gave me a good telling off. I was uncomradely and cowardly he said looking down at me sadly with his head on one side. I explained to him why I had not gone with him and why I would not go with him later. 'Then I will not go either!' he said, and neither of us went—without any apology or explanation.

Nothing happened at first. But at the next French lesson a new teacher came to take the class.

The 12th class had finished their Abitur and had been dismissed. I was friendly with many of the class and took an active interest in all the difficulties they encountered during their time of preparation. They all experienced it as a time of torture. Those who were not able to pass the 'memory test' were sent down into our class and enjoyed another year at school with us. Nevertheless, these pupils were inwardly much further advanced than we were: they had had two serious meetings with Rudolf Steiner, one in which each had been given individual career advice, the other in which he bound each one to the Waldorf School spirit and gave them the pupils' meditation. He advised them to meet together every year, and this we did until the school was closed. So many threads of life were woven together—or loosened—in these meetings. They were an opportunity for us to bring our karma into harmony.

After the Abitur at Easter 1924, our 11th class was called to a consultation with Rudolf Steiner in the library. We sat at small tables in horseshoe form facing Dr and Frau Dr Steiner and with a whole row of teachers standing behind us. We were only conscious, however, of Rudolf Steiner, Marie Steiner and ourselves. Rudolf Steiner now began to speak about the meaning of the Waldorf School curriculum. He spoke about the school's 12th year which was intended to sum up all that had been learned in the previous years. Through the Abitur itself, however, we should find ourselves squeezed like lemons—and what was squeezed out of us would first have to be put in. Through that the whole sense of the 12th class—even of the whole curriculum—would be lost, and what we should have been taught in the 12th class would be sorely missed during the rest of our lives. Then he asked us if we would agree to staying an extra year at the school, so that a 13th class could be added to prepare for Abitur. This would then be separate from the rest of the curriculum.

He did not ask the parents. He asked us. Each one of us should give his opinion. Some said they would like to stay another year at school—I knew that I would be very pleased to stay, but I was too shy to say more than that I agreed.

That was how the 13th class was introduced in 1924, so that it did not interfere with the curriculum. Doctor Steiner had hoped, however, that the Abitur would have been done away with altogether by the time I had got so far!

When I arrived at Koberwitz in 1924 Rudolf Steiner was already there and gave me a hearty greeting at breakfast. Then there was the tour of the park and gardens. The ladies were decked out in their finery. It was a splendid, festive day.

Rudolf Steiner had already noticed our brown water and wanted to know if it could be used for medical purposes. We proceeded to the purification plant where the water was filtered. The coachman turned on the tap of the tank and a rather thick brown residue spurted out in a jet. The ladies laughingly rescued their white shoes from harm and called out to Rudolf Steiner to save himself—he, however, calmly stepped across the doorway and was then on dry land. He asked for a tumbler and filled it with the brown liquid, held it up to the light and tasted some of it amidst loud protestations from the ladies, to which he paid not the slightest notice. Then he dipped his thumb and forefinger into the glass and rubbed them together to test it. Only then did he turn his attention to us and said that unfortunately there was insufficient arsenic in it.

Next we walked through the garden and park to the ponds and returned by way of the rose garden. My mother complained about the poor show of roses, but Rudolf Steiner said that nothing could be done about it as it was the result of too much iron. But my mother would not be consoled with that and protested vigorously that there must be something which could be done to make the roses blossom better. Rudolf Steiner thought for a moment and said: 'Try highly diluted lead'.

We, the young people at Koberwitz and Breslau, had taken on the task of guarding Rudolf Steiner. There was a reason for this in the fact that a plot to kill him had been discovered in Munich before he set off. We relieved each other every two hours. I had been allotted the early watch from 8 until 10 am. We carried a pistol in our pocket and all of us could shoot.

Rudolf Steiner took breakfast with his wife in my mother's drawing room. I sat in front of the door. Suddenly he came out of the room but soon went in again. Soon after that my cousin Aki came to see if everything was all right. I said to him that I had been thinking during the watch about the fact that everybody asked Rudolf Steiner questions—I would also try to ask him a few questions. Aki said: 'If you want to do that, do it now, whilst it is your watch, otherwise you will never get the chance!' So I took paper and pencil and thought awhile. There were plenty of questions: concerning Waldorf pupils, about the relationship between boys and girls, about reincarnation—I wrote down 30 questions. Then Rudolf Steiner came. It was my duty to walk at a distance of two metres behind him. Now was my chance: I had to take my courage in both hands to ask for an appointment. It took me until halfway down the stairs. Then I asked him: 'Herr Doctor, may I have a word with you?' He stopped still and asked: 'What do you want to say to me?'

And when he turned towards me I felt such a stream of love and kindness surrounding me. All anxiety about the possibility of saying something wrong left me. On the contrary, I became strong and more confident in my inner being.

Many people have later told me the same: through his gaze and through his presence he awakened one's higher self. One felt oneself protected and confirmed.

I had meant to reply that I had thirty questions to discuss. But in the inner peace of the moment the thirty-first question came to me: 'Herr Doctor, I would like to have a meditation.' 'Yes,' he said, 'with pleasure, but not now, I have to greet the guests now. Come to me during the mid-morning break.' After the lecture Rudolf Steiner went into his room and I positioned myself in the dining room. Then he came out and seemed to be looking for me. I caught up with him in the entrance hall and said: 'Here I am.' From here the broad staircase led up to the first floor. The hallway, stairs and front door were swarming with people engaged in lively conversation and eating. And amongst all that this significant moment of my life took place. I describe it here in detail in order to show how different things are in this respect in this age of the Consciousness Soul than they were formerly, when the candidate for initiation had to fast and carry out preparatory exercises to attain the right mood of soul before he could be accepted as the pupil of an initiate. For by receiving a meditation one becomes such a pupil. But here, in the midst of the bustle of civilisation, Rudolf Steiner said to me: 'First of all, of course, you need to carry out the *Rückschau*' (review of the day's events). That arose so much from the depths of his soul that I realised the retrospective review was the basis of every meditation. Then, as my meditation, he gave me the beginning of St John's Gospel for evening and morning, in connection with sunrise and sunset and in union with Christ.

I experienced that as the most beautiful meditation I could possibly have received. It led me to a study of this Gospel and to the St John path of schooling; and it no doubt also led me to the Gargano region where I rediscovered the ancient site of such a schooling.

In the course of the years I was able to ascertain that Rudolf Steiner gave this meditation especially to those who were connected with the Koberwitz impulse, because the way of St John is also the way leading to the interior of the earth.

The nine steps of this schooling enable the pupil to penetrate the nine layers of the earth to its centre, the golden Light-Centre or, as it was formerly known, the golden land of Shambhalla.

We know that when the blood of Jesus Christ flowed down onto the earth He connected Himself with the earth. Then His way led on Good Friday and Easter Saturday through hell to the centre of the earth, from where He could rise up again by permeating the earth's layers. Just as this Sun-Spirit rayed down from the cosmos onto the earth in the time before Golgotha, so now do His spirit forces ray forth from the centre of the earth into the cosmos, and everyone who unites with Christ in his consciousness and in his heart can share in that. This is described by Rudolf Steiner in his lecture *The Etherisation of the Blood.* Lazarus-John also takes this path through the nine layers of hell, and at the 7th step he is able to resurrect.

When my mother asked Rudolf Steiner during the Koberwitz Course if she would get to the golden land of Shambhalla if she were to sink down to the centre of the earth, he said to her: 'Yes, if you want to sink vertically into the depths, you will reach the centre of the earth, and it is of gold. Faust also stamps with key in hand and sinks into the depths, into the realm of the Mothers. With the powers he receives from here he is able to create Helen and also Euphorion!'

Did not Rudolf Steiner tell us in his 'Address to Young People' after the Agricultural Course, about the Altar under the earth where man is to search for the Sword of Michael to bring it up into the world?

The way of St John leads via the stage of Mystic Death to the experience of the grave, then into pre-natal existence, to the Mother, to the 'New Isis', to the Goddess of nature who can draw cosmic forces together into birth. And leading cosmic forces down to the maltreated earth is the task of the 'Koberwitz impulse'. This is the essence of Rudolf Steiner's last Whitsun message to mankind.

PART III

The Whitsuntide Gathering in its Historical Setting

Concluding remarks by
Count Adalbert von Keyserlingk

The historical stage: events between 1912 and 1924

It seems necessary to me to look at this Whitsuntide impulse in the light of Michaelic knowledge, so that its full significance may be appreciated. One should remember that in 1902 Rudolf Steiner wanted to start giving lectures about karma, but postponed this plan because of the fear people had of such knowledge. As a result, when war broke out twelve years later, its causes could not be ascertained, nor could the powers of destiny which then came into operation be recognised.

The Anthroposophical Society was founded in 1912. The Christmas Gathering took place in 1924. At the mid-point between these two events—1918—the whole fabric and structure of society altered.

These historical events were set in motion at Michaelmas 1914. They were closely connected with the destiny of the younger von Moltke who, as formulated by Rudolf Steiner, was deeply bound up with the German folk spirit. The collapse in Eastern and Central Europe led to the triumph of western materialism.

Such a way of looking at things may seem far-fetched to those who are satisfied with the generally-held opinions which have had such a hand in forming the destiny of our times. People do not want to look beyond present events because they are afraid of doing so.

But the fate of Colonel General Helmuth von Moltke, Chief of the General Staff, can teach us a lot about the Michaelic mission of our folk-karma.[1]

The forces of opposition have been trying for decades to destroy the heart of Europe, the then centre of the world, so that harmony between thinking and willing, between West and East, might not come about. Mankind was to become ill; its heart forces were to be impaired.

On 14 September 1914 the outcome of this battle was settled for many decades to come. (It was on this day that leadership of German military operations was handed over to General von Falkenhayn.) Adversarial powers were able to take hold of the centre.

The Gabrielic Age of 'Kaiser and Empire', with divine grace and hierarchic order, was past. The monarchs still had power and received recognition, but they no longer had insight into divine intentions. They destroyed themselves! It is fascinating to observe how Nicholas II, Franz Joseph and Wilhelm II destroyed their sovereignty by wrongly intervening and giving wrong orders.

In August 1914, on the first day of mobilisation, the German Kaiser gave orders to withdraw the offensive in the West and to send parts of his army to the East. His Ambassador in London, Prince Lichnovsky—a convinced Anglophile, no doubt because of his Eastern blood—had wired an incorrect message, which was presumably a result of, in this case fatal, optimism. On the strength of this communication, Wilhelm II believed he could rely on English neutrality.[2]

But as the advance had already been planned down to the 'last nail in the horses' shoes', von Moltke would not put his signature to this imperial order, which would alter his plan of campaign. When, during the same evening, Lichnovsky's communication proved incorrect, von Moltke was immediately called before the Kaiser and given a free hand once more.

When Eliza von Moltke talked about the Kaiser's behaviour later, indignation rose in her every time afresh. The drama which took place that night after the Kaiser had discovered Lichnovsky's error, was sufficient in itself to make one realise what von Moltke himself must have gone through after all the to-ing and fro-ing. He, who had sworn unconditional obedience to his supreme war lord, was fully aware of the dire consequences of such obedience in the case of false orders from on high.

How devastating these consequences actually were was shown a few weeks later. It led, as the French expressed it, to

THE HISTORICAL STAGE: EVENTS BETWEEN 1912 AND 1924 165

the 'Miracle of the Marne'. Again, acting on highest orders, von Moltke was forbidden to advance his headquarters, and therefore he sent his Lieutenant Colonel Hentsch to the Generals Kluck and Bülow, that is to the 1st and 2nd Army. In *Weltgeschichte in Bildern* ('World History in Pictures') it is stated thus:

> The General Staff sent Lieutenant Colonel Richard Hentsch with far-reaching, but only verbal, authority to the Front. This Officer, owner of majority shares in the Banque de France and a Freemason, on his own initiative ordered the cancellation of the siege of Paris. Nevertheless, the battle turned out favourably for the Germans to begin with, but on 9 September it was broken off and the Front was withdrawn. This positional reverse cost von Moltke his job.[3]

(In spite of repeated enquiries to the editor of *Weltgeschichte in Bildern*, it was not possible to ascertain the source of this information.)

The battle thus progressed in favour of Germany until 9 September, but was interrupted and the Front withdrawn. This at least is the French historical account.

Hentsch's order, given against better knowledge, created disorder, inconsistency and opposition among the war lords of the 1st, 2nd, 3rd and 4th German armies and undermined the soldiers' confidence. It not only brought gaps in the front line, but even overlapping between sections of the army. The French General Goutard writes in his book in 1953 of an 'unexploited German victory'.[4]

The English were able to break through; the German right wing was in danger; Paris was mobilised and a great French offensive was in preparation. During this chaos von Moltke ordered the retreat of the Western Front on 13 September in order to connect up his armies again, to retain the link with his reinforcements and—after having re-established order—to begin a new offensive. All war reports are agreed that this was the only possible way of directing the battle towards victory.

But the Kaiser is indignant: 'We are retreating!' On 14 September he relieves von Moltke of his supreme command

and gives it to his opponent, General von Falkenhayn, who ordered the trench warfare which von Moltke had wished to avoid.

This trench warfare allowed the army to bleed to death in inactivity. To begin with Moltke was even obliged to sign the orders of Falkenhayn in order to retain the confidence of the troops. That is one of the most difficult pieces of self-denial that an honourable person, convinced that he owes unconditional obedience to his supreme commander, has to undergo.

An essay by Walter Goetz put it like this:

> He [the Kaiser] dismisses the sick Count von Moltke at the end of September from his post as Chief of the General Staff and some weeks later appoints General von Falkenhayn to be his successor. The Prime Minister, with Hindenburg and Tirpitz, raise objection to this: Falkenhayn does not command the respect of the troops, but the Kaiser does not let that influence him.[5]

At that time von Moltke felt quite well, but could not stand being condemned to inactivity at headquarters and went to the Front—perhaps to seek death at the capture of Antwerp on 2 October.

He wrote to his wife on 3 October: 'You know that I have refused to report sick and to leave. You know that I told you that my place is here. I stand and fall with the troops.'[6]

He wrote to her on 11 October to say that he was no longer being informed by Falkenhayn of the state of the campaign and added: 'Yesterday I returned from Antwerp. There I was at least able to help a little, whereas here I am only an observer. Antwerp was at least a success again at last.'[7] He was awarded the Iron Cross for it by the Kaiser.[8]

If one envisages the torment of soul which von Moltke must have endured, one can understand how he eventually really did become ill. He saw how everything went wrong, nevertheless had to sign the orders and was not allowed to act. He was not even given any information. On 22 October, five weeks after having been dismissed, he wrote: 'My body had stood it all so well up to now, but I have finally had a collapse. It is an inflammation of the gall bladder and congestion of the right lobe of the liver. In itself the complaint is not serious, but I have to lie flat. The Kaiser's Doctor N. treats me in a nice and friendly way. He thinks that it will get better in a week.'[9]

That means it was a temporary inflammation of the gall bladder, after having plagued himself for five weeks to the point of exhaustion! What had formerly been spread abroad about his illness was therefore not true!

He was visited by the Kaiser for the last time on 23 October 1914—by the man whose confidant he had always been, who had had him on his yacht 'Hohenzollern', and on other summer excursions to Russia, and to England. Von Moltke always had to be there with him. But now, after the events of the Marne, everything was suddenly different.

The formerly invincible army of Central Europe became paralysed in trench war in the East and West. At home inner strength and moral force crumbled away.

Von Moltke, however, soon overcame his gall infection and reported himself well again. Thereupon he received a letter on 3 November giving him his final dismissal 'with the greatest regrets of the Kaiser, on account of his illness'. He returned to Berlin. The adversarial forces in conjunction with Falkenhayn—whose personal circumstances would be worth a detailed study in this context—had been victorious! It was only with difficulty that von Moltke was able to learn from Berlin what was going on at the Front. Letters to the Kaiser with suggestions went unanswered—he, like Admiral Tirpitz, had been immobilised.

On 18 June 1916 von Moltke gave a funeral address for his friend, General Field-Marshall von der Goltz, who had had similar experiences to himself. Immediately after his address and still during the funeral ceremony he collapsed and died.

Opinions about von Moltke are still so coloured by slander and distortions that it is only now possible, after nearly 60 years of thorough historical documentary research, to establish the truth, even though Jules Sauerwein published an interview with Rudolf Steiner in *Le Matin,* in October 1921 already, in which he directly questioned the latter about rumours that were being circulated:

> —If your opponents are to be believed, the Chief of General Staff first lost his head, and then the Battle of the Marne, on your account.

Rudolf Steiner's reply to that I shall here quote in part, because it describes the situation at that time:

> —I am not surprised at what you tell me. No means are spared to drive me out of Germany and, if possible, out of Switzerland. These attacks are based on a variety of causes. But in as far as my connection to von Moltke is concerned, they have a quite definite intention. They aim to hinder the publication of some notes that von Moltke made for his family before he died and which I was to have made available to the public in book form with the consent of Frau von Moltke. These memoirs should have appeared in 1919 already.

That had been made impossible by diplomatic agents who visited Eliza von Moltke and also Rudolf Steiner.

> —But from this becomes apparent what is perhaps worse, that the Imperial Government was in a state of complete disorder and was ruled by incomprehensibly thoughtless and ignorant people. One might apply what I wrote in my foreword to those responsible: 'Not what they did led to the disaster, but the whole nature of their personalities.'

Because politics were at their lowest ebb, the whole decision—one of worldwide importance—rested upon the shoulders of a single person, von Moltke.

> I never spoke with von Moltke about political and military questions before his resignation. Only later, when he was seriously ill, did he speak to me openly about all these events. In August (1914) I saw General von Moltke once only, and that was in Koblenz on 27 August. Our conversation was about purely human matters.[10]

Rudolf Steiner then describes the events leading up to August 1, 1914, which were almost unbearable for von

Moltke who bore the whole weight of responsibility. In an additional note he adds the following:

> In the first place I consider the present moment one in which everyone must speak out who knows anything about the truth of the war. I knew Herr von Moltke and with time came to value the nobleness and purity of his personality. His lips, I am sure, never uttered an untruth. He was put in a tragic position in July 1914. He knew the awfulness of the decision and his military duty forced him to make the decision alone. It is my personal view that talk about blame for the war follows a quite mistaken direction. One cannot speak of blame at all as people do. It is a case of tragedy.[11]

Von Moltke once again came to the defence of Germany when he wrote to the Prime Minister, Bethmann-Hollweg, and to the Kaiser, on 15 January 1915, to say that one would be forced to conclude peace before there could be an armistice, owing to lack of food for the people and the army. He made concrete suggestions as to how it could be put into practice and begged for immediate action. As a result he was forbidden to interfere any further. It was only months later, when it was already too late, that his suggestions were taken up. Through that too he united himself with the German people, as he had done previously in military ways.

Rudolf Steiner wrote a personal letter to von Moltke on 23 November 1915, which has kindly been put at my disposal by von Moltke's granddaughter, Frau Via Burgheiser. As this letter was not only of interest to the Chief of General Staff, I shall here quote some of it.

Rudolf Steiner wrote somewhat as follows:

> Your Excellency, you are passing through a time in which your destiny is becoming a riddle in your life. In these paths of destiny the powers hold sway which oversee the spiritual guidance of man. Your spiritual guides have woven you with your own inner destiny into the leadership of the German nation in this epoch. This destiny of the German nation is bound up with the deepest and most exalted goals of human

evolution. The threads of such a national destiny as this are not simple. They must often become entangled. The way of destiny proceeds through trials, trials which lead to the edge of the abyss of world secrets. To the abyss, where the great question 'To be or not to be' arises before the soul, where apparent darkness spreads out before one's eyes. But for the German people there stands at the abyss no spirit with a lowered torch, but one with a torch held high. The way towards the light will be found, whatever happens. Hindrances and difficulties will only mean that powers will grow in the face of them, so that the way can be found and the genius of the nation followed.

Whoever, like you, Your Excellency, is bound to this folk-destiny, for him the destiny of the nation is mirrored in his own destiny. The spiritual leadership of mankind has given you the obligation to lead the German nation to a phase of its task. The fact that you have come, at a certain moment, to a seeming halt, is only so that you can gather new strength. You were, in truth, always prepared to devote your forces to the events of the time. Things are in reality quite different from the way they are presented to the outer senses. You are not now outwardly connected to the scene where events are being enacted; but in the depths of your soul you are connected, and in these depths the forces you will need when the powers of destiny call you once again, are collected together and grow stronger. And destiny needs these forces of yours. It prepares them within you so that it can make use of them in the right place. Your way is the one all must tread who have a serious task to perform. This can be summed up in the words: To await peacefully what is to come, and to be prepared when one receives the call.

One can therefore say that on 14 September 1914, in the first days of Michaelmas, the old order 'toppled itself' when the Kaiser withdrew his support from von Moltke and gave it to his opponents. German development from then on took a disorderly, chaotic course. It entered into a time of riot and collapse, starvation and senseless wastefulness. Impulses pressed in from East and West, seeking to destroy Germany. And, unbelievably, the German leadership itself also contributed to this. Radical forces in the East were given financial support in order to bring about the overthrow of the Tsar, and Lenin was sent to Petrograd.

It is quite clear that adversarial forces lost no time in making use of the shortsightedness and limitations of the German leadership to further their own ends.

Inside Germany Socialists and Nationalists fought one another and were incited to it by East and West. What Germany intended to do to Russia happened to itself, and this grew to a world-shattering state of conflict.

A figure who seems to me to be a counterpart to Helmuth von Moltke was Lenin, who sought to raise the active forces of this time of inner change into tangible expression in the outer world situation.

He was born in 1870, while his confederates Trotsky and Stalin were born in 1879—the year in which Michael finally exiled Ahrimanic forces to the earth.

Already in 1887 Lenin was expelled from Kasan University for causing political unrest. Some years later he was sent to Siberia. In 1900, while in exile, he founded the underground paper *Iskra* in Leipzig. In Russia the newspaper *Pravda* appeared in 1912, though at first only sporadically. Thus was formed the basis for a complete change of the previous social order into its opposite.

The Tsar, descended from the Romanovs, inheritor of absolute power over the life and death of all his subjects, was replaced by persons of unknown name, who had first to assume new names for the roles they had to play on the stage of history: Uljanov called himself Lenin, Dschugaschwili called himself Stalin and Bronstein called himself Trotsky. They took over the same authoritative position in the state, but destroyed everything else belonging to tradition.

Lenin, from his youth on, was a slave to his ideas (as were later Mussolini and Hitler). He was banished to Siberia for his beliefs, he starved, lived inconspicuously in hiding in exile, and for years lived with a cobbler in Zurich—but he never ceased his study of revolution and its inspirers and opponents, whether in libraries or in conversation. For decades on end he pieced together a detailed plan of campaign which he constantly improved. His ideas encompassed the globe like a gigantic net. His companions in many countries were at work under his influence, to bring to realisation his idea of a completely materialistic state of Russia, without God. He was the exponent of those forces which, prematurely and in a mistaken way, seek to establish a purely physical paradise, one where all are treated the same and where no account is taken of differences in spiritual development.

In 1917 the still unconquered German army was threatened by danger from the USA.

President Wilson wished to destroy the old Central Europe with ideas that, while seemingly brilliant, were nevertheless unrelated to actual realities and needs, since they came from a foreign, western source. Although he was unable to grasp the situation in Europe, he interfered in a thoroughly unjustified manner. The incompetent German Command was at a loss to know how to respond and, with an inconceivable lack of responsibility, gladly accepted Lenin's demand that he should be sent to Russia, because he had promised to conclude a separate peace treaty, against the will of the Western forces, immediately he had seized power.

Stefan Zweig described how the 'sealed wagon' quietly set off on the night of 12 April with Lenin and other Russian emigrants on board, and with a German guard, through Germany to Sweden; and how Lenin was received with pomp in Petrograd on 16 April by comrades, functionaries and workers, and the militia, as their acclaimed leader. The revolution now enters a new era. A telegram on 21 April 1917 from the German High Command to the Foreign Office reads: 'Steinwachs [Leader of German military intelligence in Stockholm] wired from Stockholm on 17 April 1917: *Lenin's entry into Russia a success. He is acting entirely according to our wishes. Therefore a cry of anger from the Stockholm "Entente Social Democrats"*.'[12]

In addition the German government supported the revolution and the editor of *Pravda* to the tune of 3 to 6 million Marks per month.[13]

The Tsar had already abdicated on 15 March 1917, three weeks after the beginning of the February uprising and was shot with his family on 16 July 1918. Only the youngest daughter, Anastasia, survived as a nameless orphan and was forced to live an inhumanely difficult life, as though the destiny of the Romanovs continued to rest upon her. This led to one of the most extensive lawsuits in Europe, lasting 55 years, which ultimately failed to restore her name to her, or recognise her descent. Many people with a destiny of this kind—similar to that of Kaspar Hauser—were born after 1912.

Lenin only really came to full power after uniting with what Trotsky had absorbed from the West, from America, and what Stalin had brought with him from the Caucasus. Although the revolution had begun in Russia in February 1917, Lenin was the first to succeed, through his brilliant oratory and journalism, in swaying the masses and getting them to rise up; and afterwards in forcing them to comply with his wishes through the most dreadful terrorism by the *Tscheka,* the secret police of Comrade Felix Dscherschinsky, who himself remained in the background. It was he who created fear and spread it throughout Russia.

It took months—from the end of November 1917 till 3 March 1918—until the peace of Brest-Litowsk could be signed. The State Secretary, von Kühlmann, General Hoffmann, and Foreign Minister Count Czernin for Austria, carried out the negotiations. The German army had not been defeated and was stationed not far away from Petrograd. The Russians were inwardly and outwardly unable to fight on. Von Kühlmann, a school friend of Count Lerchenfeld of Koefering, was able, through the mediation of the latter, to speak with Rudolf Steiner about the necessity of peace terms in the East which could be guaranteed to hold. They also spoke about the Threefold Commonwealth. In spite of that Kühlmann made peace conditions which did not take account of the essential needs of other nations. As Trotsky and the Russian Commission expressed it, Russia was to be deprived of a territory equal in size to that of 18 provinces.

When the bearer of the Treaty documents was about to spread them out before Lenin, the latter remonstrated: 'What! Must I not only sign this impudent document, but also read it too? No, never! I shall neither read it, nor carry out its articles on any occasion where there is an opportunity to do so.'[14]

Two years later Lenin said: 'Germany only got a couple of million pounds of corn out of the Brest-Litovsk peace treaty, but along with it the subversive germs of Bolshevism. But we have gained time—time in which the nucleus of the Red Army could develop.'[15] How right Lenin was when demonstrated by the Treaty of Versailles and later by the Second World War.

Lenin directed Bolshevist-Communist development and everything else. He was absolute dictator. His plans for the electrification of Russia were already realised by 1920. In 1923 a new constitution was decided upon. Lenin died on 24 January 1924—in the same year that the Agricultural Course was given. His post-mortem examination, at which Lenin's GP—my teacher, Professor Foerster from Breslau—was present, testified to the fact that Lenin's brain had shrunk in its whole mass to about a quarter of the normal volume. 'When we opened it up,' writes professor Rosanov, 'we found an extensive sclerosis of the brain vessels—and nothing but sclerosis. The most astonishing thing about it was not that the capacity to think remained intact in such a calcified brain, but that he was able to remain alive at all.'[16]

One has to ask: who was it who did the thinking for him, in order to bring Bolshevism into existence?

After Lenin's death, his political will was not observed. His friends were dismissed—like Trotsky, who was banished in 1927 and murdered in Mexico in 1940.

At the time of the Agricultural Course, the severest power-struggles were taking place place in Russia, both openly and in secret, from which the Caucasian Stalin emerged victorious in 1925. The powers coming from the East had been able to free themselves from western influences. They had become strong enough on their own.

THE HISTORICAL STAGE: EVENTS BETWEEN 1912 AND 1924 175

Let us now look at another group of people, also intent on using for its own ends the new forces which are flowing into the stream of earthly life—the group, that is, striving not to create a world revolution but to build up its own independent national individuality. The people in this group strive to set geographic limits and enclose their own might and splendour within their nationality. Wilson's demand for the right of self-determination of nations opened up the way for this and gave it the appearance of legality.

Mussolini (1883–1945), as also Hitler (1889–1945), succumbed to these ideas in early youth. Mussolini, born the son of a blacksmith, seemed to absorb raw iron and Mars-forces into himself even in childhood, which then continued to work on in him. First as an unruly schoolboy, then as an incompetent teacher, he championed revolutionary ideas with wild enthusiasm in writings and in speeches and turned against every kind of discipline. He finally had to go to Switzerland, where he worked as a bricklayer and came into touch with Russian emigrants. After a period serving in the army in Italy, he worked on the paper he founded, *The Class Struggle,* and in 1912 he wrote: 'We wish to show that our Fatherland exists as little as God does.'[17]

Surprisingly enough, however, a remarkable change came over Mussolini in August 1914. He was excluded from the Socialist Party, founded the paper *Popolo d'Italia*, and in October 1914 came out in favour of the war against Germany. The declaration of war on 24 May 1915 filled him with enthusiasm. He became a volunteer Lance Corporal and was later seriously wounded. During the post-war chaos he founded his Fascist Party with discharged troops, which, by 1919, already had 17,000 adherents. In 1921 it took over the task of establishing order in the country and created new work-places. In the meantime, the number of followers had risen to three quarters of a million, and at the election on 15 May 1921 Mussolini won 35 seats in parliament. Everwhere he was regarded as a saviour.

On 18 October 1922 he declared war on the official government and from 27 to 29 October 1922 the March on

Rome took place. Now the man of the moment is joined by the Court and Church. Mussolini is the people's hero. He becomes the head of the Government, 'Il Duce'.

He regarded Garibaldi as his folk-hero, whose ideal it had also been—in spite of the Risorgimento—to acknowledge the royal family. Italian Fascism consolidated between 1914 and 1923/24—in other words also at this time of the outer and inner reshaping of society. This system was then taken by many states as a model for their own regimes.

Adolf Hitler felt an inner summons already in 1906, when he was 17, and developed from it his detailed plans for social reform so as to create a future 'ideal' state of Germany. From 1907 on he turned his revolutionary plans against all ruling parties and studied intensely in libraries until, in 1908, he found Gustave le Bon's book *Psychologie der Massen* ('Psychology of the Masses') which he took as the basis for *Mein Kampf*. In 1913 he went to Munich, enrolled as a volunteer after the war had begun, and was awarded the Iron Cross as Lance Corporal in 1918.

Towards the end of the war he suffered a gas attack, and after being unconscious for a long time he became a 'completely changed' personality. As member No 7 of the German Workers' Party he intensified his activities from 1919 onwards, which took him from a smoky back-room to giving grand speeches to the masses. His words had the same magical effect on the public as Lenin's and Mussolini's and those of other dictators.

Hitler was imprisoned after the unsuccessful attempt at a putsch in November 1923 and fate sent him nine months in which to write his book *Mein Kampf*. His political era started after his release from prison.

At the same time disturbances and revolutions were taking place in China and India. The cultures which had lasted for millennia were breaking up and bringing new impulses to the surface: in 1912 the Chinese Emperor was forced to abdicate. The National Party, led by Sun Yat Sen, gained victory in a bloody civil war against the Emperor's armies. After the death of Sun Yat Sen in 1925, Chiang Kai-Shek, as leader of the Military School in Peking, clamped down on the Communist movement. He ruled as absolute dictator, loved by the people and marvelled at by the world. Between 1930 and 1934 he opposed Mao Tse-t'ung, who had been active ever since the founding of the Communist Party in 1921. There had been bloody conflicts among the students in 1919 already—the 4 May movement.

Out of the Communist Party, founded by Professors from Peking University, Mao Tse-t'ung soon formed a peasant movement, which ultimately led, on 1 October 1949, to the People's Republic of China.

Gandhi, who was born in the same year as Rudolf Steiner, returned from Africa to India in 1914. He took up passive resistance against British rule in 1920. But, against his wishes, disturbances broke out in 1922, and he was put in prison in 1924. From then onwards he devoted himself to getting rid of the caste system which had existed for millennia.

In the case of Gandhi one has the impression that his struggle, by contrast with that of others, proceeded from wisdom-filled insight into the needs of the time, quite without violence. It led in 1947 to a regulated freedom under a quite new 'modern' regime of political and social community life. But only a year later, Gandhi was murdered.

In Turkey, too, thorough change was achieved by outstanding ability and equally penetrating insight into the needs of the time. Kemal Ataturk (1880–1938) had been filled from his youth with the deepest enthusiasm for the idea of Turkish reform and national freedom. He always represented perfection—which is what *Kemal* means in Turkish: this was the name he was usually given by his teachers, both in his private life and as an officer in the First World War in which he served under General von der Goltz. When the war ended he gathered troops together in the Anatolian mountains, against the Sultan's orders, and called various Pan-Turkish Congresses together to protect his people from the unrealistic conditions of the Paris Treaties, and those of Wilson.

On 3 May 1920 he formed a provisional government and, as Marshal, reconquered Smyrna from the Russians.

The Republic was proclaimed and Ataturk, as 'Father of the Turks', was its first President. Sultan Mohammed VI had already been deposed in 1922 and the Caliphate, along with Ottoman Theocracy, was abolished. In this case, too, centuries-old traditions were replaced by new political and social orders.

Similar observations can be made in Central and South America.

In Eastern Europe, Poland, in the immediate vicinity of Silesia, was proclaimed a separate kingdom again on 5 November 1916. Its boundaries were not only confirmed by the Versailles Treaty, but claims were raised in respect of West Prussia and Upper Silesia. Marshal Pilsudsky, who was realistic, directed his activity more towards the East, whereas the fanatical Dmovsky turned his attention towards Western interests and demanded large tracts of Germany.[18]

Wilson said the following in his speech to Congress on 8 January 1918:

> An independent Polish State should be set up which should include the territories of undisputed Polish inhabitants, to whom a free and safe access to the sea shall be assured and its political and economic independence and territorial immunity guaranteed.[19]

Thus the surrender of the 'Corridor' came about, and also voting in Upper Silesia, after Marshal Pilsudsky was confirmed as Head of State in Poland on 14 November 1918. Poland annexed these German territories with the support of the Western powers, against the wishes of the inhabitants of these German districts.

A violent struggle began in Upper Silesia, which concerned the vital mineral resources of coal and metal ore. The commission set up by the Allies under the polling superintendent Korfanty, formed a provisional Polish Government long before the voting began. Troops were employed to keep order, but failed to do so.

The terror of the Poles and the behaviour of the French Commission showed that it was out of the question to expect an orderly or true vote. Though thousands of Poles and 150,000 Germans travelled to their homeland to vote, it was nowhere possible to implement the policy which the French had also advocated in the League of Nations. The Poles wanted to weaken, lessen and reduce Germany in all spheres, in order to put its 'dangerous' neighbour out of action. It was a policy which was bound to lead to national uprising and new wars, as Rudolf Steiner again and again foretold. Nevertheless, of those in Silesia who voted on 20 March 1921, about 60% were in favour of Germany.

As the Poles wanted to annul the results of the vote, the third great uprising of Polish insurgents against Germany broke out on 3 May 1921. Since the Allies, under the influence of the French, did not take any action, the Germans resorted to self-protection and defence of their borders, and under General Hoefer a successful storming of the Annaberg, the holy Silesian mountain, came about with the help of the Erhardt Brigade on 21 May.

The League of Nations decided, counter to the result of the vote on 20 March, to give the Poles 30% of the contested area, together with 80% of the industrial area of Upper Silesia. Through that, too, the basis for a new war was created, which was then unleashed by Hitler after the feigned attack on the Gleiwitz transmitting station in 1939.

Rudolf Steiner had repeatedly warned about the voting: one should refuse to vote, he said, but should try to create an independent area capable of forming an economically free 'nucleus' according to threefold social principles. My friend Walter Kühne and others risked their lives, according to what Emanuel Vögele told me, to convince the people of Silesia about this new solution—even in front of mass gatherings of 3,000 or 4,000 people. They lived in Koberwitz and their activities—which extended far into Upper Silesian territory—were financed from there.

What would it have meant for the threefold structuring of Europe if, instead of becoming the centre of caprice and discontent, it had become the nucleus of a new social order, for which a majority of the population was in favour.

It would be possible to extend these reflections further—and inner amazement at all the things which have happened between 1912 and 1924 would continue to grow. Since the end of the 19th century the old order has become unhinged and something quite new has started to take powerful effect. The transformation of mankind's soul forces is being misused on all sides, and what is old, venerable and hierarchically ordered is being utterly eradicated. But new seeds are sprouting in the churned up and still chaotic ground of human destiny.

Such a picture has already been used once before as a sign and symbol of the development of the German people—in the shape of the broken 'Irminsul' on the Extern Stones.

It is as though cosmic powers send two kinds of representatives to the earth—the destroyers and the sowers of new seeds.

The individual destinies of dictators strike one as being very similar: all had a definite impulse from their childhood. In spite of suffering the greatest privations, imprisonment and banishment, they developed their own ideas through tremendous diligence and purposeful application—whether for world revolution, or for the nationalistic promotion of individual nations. They build up an illusory world, but one that is either premature or resting on outworn impulses, or on a wrong tack—they destroy what exists already, but what they build up has no stability.

In the midst of such chaos stands Rudolf Steiner. He does not, unlike the others, exert pressure on the masses, through promises, fear, money and magical speeches. He can only work on individuals by means of the spiritual truth. He may not force people by magical means, nor snatch at power by the use of weapons, or by instilling fear into them. His rules are those of freedom, conviction and knowledge.

He knows more than all the others; he could, if he so wished, work more powerfully on the masses than all the orators and national leaders. But he is obliged to wait, to see if humanity will accept what is offered in the way of urgently needed wisdom imparted by the folk and Archai Spirits for our present era. He himself writes on this theme:

> We would, without any doubt, make good progress with the working classes if party chiefs did not put so much effort and energy into pulling the carpet from under us; and in this respect the proletariat obey them more submissively than any Catholic does the Head of the Church. And the bourgeoisie, as a class, snore away in blithe ignorance. They let themselves be roused now and again to give 'explanations', and remain in the hands of those who really pull the strings, whose only knowledge of effective methods is of those which are opposed to the spirit. The fruits they produce together will be bad ones.'[20]

Into the chaos of personal thoughts and intentions, Rudolf Steiner placed objective, Michaelic thoughts, which could throw light on the pathways of humanity if they were to be accepted. But the majority of people are unable to open their hearts to receive these thoughts. In the last words which Rudolf Steiner directed to his friends, he said: 'If I had only four times twelve people, the world could be saved.'

Although the 'leading lights' of the day knew about Rudolf Steiner, and although it was the destiny of each one of them to meet Rudolf Steiner himself or those who represented him, successful realisation of these ideas did not take place in the political and social realm!

One gets the impression that many destinies unfolded in such a way that Rudolf Steiner's helpers could gather around him, to support the tasks given them by their preparation in the Michael School. Among them were some who had incarnated too early or too late, or who had other tasks to perform, but whose karma was changed by their wish to be a contemporary of Rudolf Steiner on earth. One can sense this if one visualises how the destinies of individuals affected the world and the destinies of nations between 1912 and 1924.

Karmic connections of the 4th and 9th Centuries

If I now follow the advice of Rudolf Steiner and search in the 4th and 9th centuries for causes underlying events between 1912 and 1924, I do it in order to awaken consciousness for such things. I am not presenting new historic facts, such as I discovered in my work connected with Monte Gargano, or as W.J. Stein did in his book *World History in the Light of the Holy Grail* and which he expounded to us so impressively in his classes in the Waldorf School. What I am trying to do, rather, is to look at history from a different angle. Our anthroposophical historians (Emil Bock, Rudolf Meyer, Reinhard Wagner) have given us so much material about the period around 869 that it cannot be very difficult to give a summary of it which accords with Michael wisdom. Here, however, space will not allow this, and therefore, as I assume that the accounts in Rudolf Steiner's karma lectures of 1924 are known to you, I shall restrict myself to recalling those times.

Actually such a view of past times should begin with the Dodona Mysteries from which clairvoyant priests of the cult of the Dove (symbol of the spirit) and the Oak tree, (the Mars-knowledge of the Druids) gathered their wisdom. It was from the name of their priests ('Helle') that the Greeks were called 'Hellenes' and their priestess, Helena, received her mystery knowledge from there. Helena, the initiate of the Dodonic 'heavenly Oracle' was stolen and taken away to Troy, and was initiated into the Mysteries of the Palladium, the Chthonic wisdom of the greatest sun-jewel, with its thirteen subterranean altars. She united with Paris, from the race of the Priest-Kings of Troy. He no longer gave the apple of the wisdom of Jupiter to Hera, the guardian of the ancient rules, nor to Athene, the representative of cosmic intelligence, but to Venus, the Goddess of love and beauty, and thereby inaugurated the Fourth Cultural Epoch. The time thus came for Troy, the seat of Oriental wisdom, to decline, and Greece to take over the tasks of the new age.

Yet the Mystery of Golgotha was to take place at the time of Rome's world domination, which is why Aeneas, the Son of Venus, had to bring the Palladium to Rome. The first Christians, too, fetch this power, Christianity, from their Mystery Centres beneath the Palladium, to distribute it throughout the world. The Palladium is the Roman relic par excellence, which lends the city its position of world-rulership and its central importance.

Constantine the Great made early Christianity, with its enormous inner power, into the state religion. He made ready the Council of Nicaea (325) for the formerly suppressed and impoverished bishops, gave them great possessions and enjoined them to affirm in the world the Roman impulse of dominion through power. He thought that he could thereby extend his influence towards the East through his city of Constantinople. In his personal lust for power he brought the Palladium to Constantinople with great pomp, and preserved it on subterranean altars beneath a porphyry pillar, upon which he placed his own bust in the likeness of Apollo, with a circlet of rays made from the nails and wood of the Cross of Christ. The pillar was destroyed by lightning. He built twelve Churches dedicated to St Michael in order to place the Hagia Sophia, the Goddess of Byzantium, under Michael's protection. The Hagia Sophia is the Pure Virgin, the Theocrata or Divine Mother, and was called Moira, or Mira. All this was done by Constantine without him being a Christian. It was only when he was on his deathbed (337), that he allowed Bishop Eusebius to baptise him, because of his fear of death. As a baptismal gift he stipulated that in his city thenceforth the Patron Saint should no longer be the Virgin Sophia (Mira), but Mary the Mother of God. Hence arises the confusion about Maria as the Mother of God which still needs to be corrected in the Michael School of Alcuin and many of the dogmas of the Church Councils.

During this 4th century Michael, the Hebrew folk-spirit

who prepared the birth of Christ and then became the guardian spirit of Byzantium, entered the West like a counter-current in spiritual history. He established his centre in Monte Gargano, the holy place of the ancient Dodonic Oracle in Southern Italy, where esoteric, Johannine Christianity could be continued. Thither followed the artists and their works of art, driven out by the image-breakers, and it was here that the worship of the Madonna found its way to the West. The German people came to meet Michael, to encounter him and unite themselves with their future folk-spirit—so did Alarick, the West-Goths, the Langobards, the Normans, almost all the German Emperors, as well as Ludwig the German. The migration from North to South, to the Michael Sanctuary of Monte Gargano, was caused by the earth itself: when the ground-level in the north-eastern Urals rose, the Mongols had to emigrate, thus beginning a general migration.

Two worlds emerge in Italy in the 4th/5th centuries, divided by the Apennines. On one side stands Rome with its later strife between Emperor and Pope; on the other side stands Michael in his battle between esoteric and exoteric Christianity. To the South-West lies Caltabelotta, the centre for Klingsor, the opponent of the Grail, where Duke Ludolf of Capua and Richard of Azzera, both with large retinues, were living. In the East rose Monte Gargano, the place whose age-old Mysteries gave rise to the spirituality of the West and where the centre of Europe evolved. Friedrich II, the son of Apulien, together with the 'Deutschritter' [an order of knights from East Prussia], lend outer form to this central Europe.[21]

There too was held probably the greatest council ever held on earth: it took place in 333 between the four greatest leaders of the earth, who decided on the direction of earth-evolution up to our own day, as Rudolf Steiner indicated in his karma lectures.

The counterpart of Michael-Grail against Rome-Klingsor which developed in the 4th century, then produced a decisive result in the 9th century which can be summarised as follows:

Charlemagne, a descendant of the Grail lineage, Alcuin, the Irish-Scottish priest, as well as the Western Mystery Centre at Niedermünster at the foot of the Odilian Mountain in Alsace, to which the Count of Reichenau brought the Grail insignia from Corsica, stand in opposition to the conquering advance of the Arabs, Saracens and Turks and the high culture of Harun al Raschid in Bagdad. The activity of Klingsor in Sicily and the personal power-lust of individual bishops within the church, had a great impact on the hearts and minds of the Christians.

Within this storm of history stands a personality, almost that of an Archangel himself—'like an Elijah', says the Chronicler Reginald of Prüm—in the midst of the waves of the opposing impulses of this decisive period—Pope Nicholas I.

The cosmic intelligence, which Michael had formerly guarded within the sun, sank to earth into human beings and was given into Ahriman's charge to help human beings make their own decisions in the freedom of knowledge.

Mankind had to sever itself from its spiritual guidance and turn completely towards material knowledge of the laws of nature. It is only the being of Ahriman who is able to bring that about. Therefore mankind also had to turn away from the cosmic wisdom of the East. For the same reason it also had to renounce the laws of karma, in order to forget its own spiritual nature which advances from one incarnation to the next. But at the same time, since 1879, human souls are being prepared to decide, in the New Michael Age, whether they want to follow Michael or to follow a further course downwards into the sub-human realm.

Pope Nicholas I and his Counsellor Anastasius Bibliothecarius clearly recognised the turning-point of time as a preparation for the Michael Age. They cannot and do not wish to check the development sent by world karma, but they try to guide it along the right lines. So Nicholas I brings about the split from the Eastern Church in a mighty battle against the Patriarch Photius, whose knowledge and brilliant oratory, based on his use of egoistic and magical forces, surpass that of Anastasius.

The Pseudo–isidoric decrees, which W.J. Stein describes as having been written by Ahriman, are laid before Nicholas; and he is unable to fully grasp that they are a forgery by the Klingsor folk in Capua. The most difficult opponent to deal with is Arsenius, the spirit of deceit, because he appears as the faithful servant of the Pope. He had one of the highest positions of honour and was the Pope's delegate to Ludwig the German. The Benedictines of Monte Cassino referred to him as a 'devil in human form'.[22] He promised to do everything according to the Pope's intentions and always managed to achieve the opposite. When his son abducted the daughter of Pope Hadrian and later killed her, he fled to Benevent, where he presently died of a mysterious fever. One can even assume that Arsenius was behind this murder.

Walter Johannes Stein writes about Nicholas I: 'The question confronting Nicholas was not: Shall this spirit of untruth be allowed entrance or not? But only: As he *is* entering our life, what direction can be given him in the soul of Central European people?'[23] He backed this up with two letters which Eliza von Moltke had received from Rudolf Steiner and put at his disposal. All these karmic events then came to their conclusion in a threefold way in 869.

As a result of the split between the Eastern and Western Church, the Eighth Ecumenical Council, described so brilliantly by Reinhard Wagner,[24] took place in the Hagia-Sophia, the church of the Holy Spirit.

The trichotomy of the human being was declared heretical and its teaching forbidden. Man consists—according to the Church, since 869—only of body and soul. Rome thus separated itself from the view of the Eastern Church, that man also consists of a third principle, the spirit. The spirit may no longer be cultivated in the Roman Catholic Church. Rudolf Steiner calls this a 'second crucifixion of Christ'. It is the most terrible occurrence to have befallen mankind on earth until then. The results for science, art, philosophy, as also for everyday life, can still be experienced everywhere in our daily existence.

It was Anastasius Bibliothecarius of all people who, as a guest at the Council—Nicholas I had died already in 867—had to take his personal notes and minutes of the meeting to the Vatican, as the official acts had been destroyed by pirates.

Rudolf Steiner draws our attention to yet another event of 869. It is the meeting of the Grail current with the Arthurian current in the South of France. The Grail current seeks Christ in the blood and hearts of men—each must do that for himself; the Cosmic Christ, on the other hand, works through Michael on the earth. These two currents meet each other. The Christ who lives within man and the Christ who works from the sun meet one other!

As Rudolf Steiner tells us, a council took place in the spiritual world, at which the souls of Harun al Raschid and his counsellor met the souls of Aristotle and Alexander. The decision of the Arabian souls to introduce the false idols of materialism into the world cannot be altered. This fact works on in personalities reincarnated in the 18th/19th/20th centuries and will continue to have consequences for the further history of mankind until far into the future.

The year 869, a turning-point on three levels, determined the future evolution of mankind. The change which occurred in human beings and in their way of life between 1912 and 1924 is a result of this turning-point in 869. Nicholas I, his Counsellor Anastasius, Scotus Erigena, and many others, were not only of decisive importance for their epoch, but have set the direction for human evolution since the 9th century. The tremendous decision to continue with a human evolution devoid of knowledge of the spirit, weighed upon them in every impulse of their souls, but the light from Mount Odile in Alsace illumined them. There seems to have been a close connection between the cloister (of St Odile) and the Pope, as I was informed by a Theology Professor whom I met there.

In 1914, however, these personalities are again engaged in a mutual battle against the 'Arabian souls' and against

Klingsor's agent, but especially also against those of Capua whose wish it is to hinder spiritual progress, and thus bring catastrophe to mankind in times to come.

To end I would like to quote a passage from a lecture by Rudolf Steiner given in Torquay on 21 August 1924, where he speaks about this period and about us today:

> In the 9th Century people were already beginning, as the forerunners of those who came later, to unfold their own personal intelligence: intelligence begins to take root within the souls of individuals. And looking down from the sun to the earth, Michael and his hosts could say: What we have guarded through aeons of time has fallen away from us, has streamed downwards and is now to be found within the souls of human beings on earth.[25]

The Koberwitz impulse as hope for the future

What took place in Germany during those 12 years between 1912 and 1924? How did Rudolf Steiner intervene to make the impulse from the spiritual world clear to people in a way which they could take up in order to protect humanity's further evolution from catastrophe?

'But they did not take it up' you may say.

I experienced this time intensively, as I was in Stuttgart from 1919 onwards. That is why I think it is so important to recall it with documentary evidence, because the causes can be sought there, not just of the Second World War, but also of what happened later. The reports from the trusteeship of Rudolf Steiner's estate (*Nachlassverwaltung*) are particularly impressive in this respect.[26]

Surrounding Germany—the nation which has the central, world-task of a particular development of the ego during the Fifth Post-Atlantean Cultural Epoch—were enemies: enemies who, since 1879, have tried to hinder the fulfilment of this task by occult and other means, in thoughts, by words and in writings.

The leading politicians had no connection any more with their folk-spirit—let alone with Michael, the spirit of the Epoch. They formed decisions out of their own ideas, but these were arbitrary and unreal. They caused rivers of blood to flow and future misery.

The one whom the German folk-spirit could work through, Helmut von Moltke, was dismissed. Possible chances for peace between 1915 and 1917 were not taken up by England, Russia or the U.S.A. Then full-scale submarine warfare started and Wilson declared war on behalf of the United States.

By his '14 points' and the Versailles Treaty, he forced the German nation into absolute submission, which created a basis for Hitler to rise to power. This in turn led to the Second

World War and the complete destruction of all the old ways of life and customs. How differently everything would have worked out if people had accepted Rudolf Steiner's suggestions.

But one must study this tragic event more closely to begin to understand the time between 1912 and 1924. Helmuth von Moltke, with his connection to the German folk-spirit, was able, at the outbreak of war, to hold out against the Kaiser to begin with. But then his enemies triumphed after all. Lies were told: 'He is ill'. After his departure everything remained paralysed. Europe was inert—a ghastly lull before the actual storm. How different it would have been if the German folk-spirit in Moltke could have asserted itself and the opportunities for peace grasped. But the spiritual world can only intervene where there is movement, not where there is rigidity.

On 13 September 1914, at the time von Moltke was finally dismissed and when his old opponents, who plunged the German nation into the abyss, took over leadership, Rudolf Steiner urged those who shared responsibility with him, the members of the Anthroposophical Society, to connect themselves with their folk-spirit and those who were sacrificing themselves at the Front.[27]

He showed how one could learn to help them by secret inner dialogue with the spirit of the nation to which one belongs. He said:

> I can only give you advice if you find a few minutes to make use of the following words, in order to find your bearings in the present world situation:
>
>> Spirit of mine earthly habitation!
>> Reveal the Light of thine Age
>> To the Christ-endowed soul,
>> That striving I may find thee
>> In the choirs of the Spheres of Peace
>> Singing the glory and the power
>> Of human hearts devoted to the Christ.
>
> (Translated by George and Mary Adams in *Verses and Meditations*)

THE KOBERWITZ IMPULSE AS HOPE FOR THE FUTURE 191

> Du meines Erdenraumes Geist,
> Enthülle deines Alters Licht
> Der Christ-begabten Seele,
> Dass strebend sie finden kann
> Im Chor der Friedenssphären
> Dich, tönend von Lob und Macht
> Des Christ-ergebnen Menschensinns.

When I met Frau Eliza von Moltke in 1922 she gave me another mantric verse for the purpose of uniting myself with the German folk-spirit—she had already given me one such meditation previously. As I learned much later, that was a meditation which Rudolf Steiner had given to Helmuth von Moltke during a conversation in 1914.

> Victorious will be the power
> For which in course of time
> The people are predestined,
> That safe in spirit care
> For healing of mankind
> They can from Europe's heart
> Light from the battle win.

> Siegen wird die Kraft
> Die vom Zeitgeschick
> Vorbestimmt dem Volk,
> Das im Geistesschutz
> Für der Menschheit Heil
> Aus Europens Herz
> Licht dem Kampf entringt.

In 1915 Rudolf Steiner again urged anthroposophists to connect themselves with the German folk-spirit and gave this meditation for it:

> The German spirit has not yet accomplished
> Its task in earthly evolution
> It lives in hope with future cares.
> It hopes in life for future deeds.
> In depths of soul it strongly feels
> What, hidden, must grow ripe for work—
> How dare, in enmity, uncomprehending,

The wish for its annihilation rise,
As long as life still manifests within
And keeps it active in its inmost core.

Der deutsche Geist hat nicht vollendet,
Was er im Weltenwerden schaffen soll.
Er lebt in Zukunftssorgen hoffnungsvoll.
Er hofft auf Zukunftstaten lebensvoll.
In seines Wesens Tiefen fühlt er mächtig
Verborgenes, das noch reifend wirken muss—
Wie darf in Feindesmacht verständnislos
Der Wunsch nach seinem Ende sich beleben,
Solang das Leben sich ihm offenbart,
Das ihn in Wesenswurzeln schaffend hält!

Rudolf Steiner started to plan the building of the Goetheanum even before 1912. It was supposed to have been completed by 1 August 1914. People of 14 different nationalities were able to work together on the wooden building with understanding and in amity, whilst the rest of mankind were in conflict with one another throughout the world.

In 1916, after von Moltke's death, events were heading quickly towards catastrophe in Central Europe. Rasputin, who had a close connection with the Russian folk-spirit, was murdered by Prince Jussupov on 30 December 1916. The Tsar abdicated on 15 March 1917 and was shot on 16 July 1918. But the world did not wake up, instead all slept on in their own egoism—yes, German politicians even supported the dangerous development which I have described, and were glad to be allowed to dictate the terms of the Peace Treaty of Brest-Litovsk.

On 22 January 1917, already, President Wilson had drawn up his American 'peace principles for humanity', which, if one reads them today, seem like the fantasies of a young man devoid of any knowledge of European affairs.

Rudolf Steiner had been making preparations for the last 35 years to promulgate the ideas and requirements that would enable human evolution to be guided further according to the intentions of the spirit. In three different ways he attempted to convey to people what he had experienced through his connection to Michael and the German folk-spirit. In January 1917 he began with his depiction of the Representative of Man between Ahriman and Lucifer. He first made a model of it—then in May 1917 he started to carve it in wood, beginning with the bound figure of Ahriman.

At the funeral service for Helmuth von Moltke in the branch-meeting in Berlin, he expounded his scientific ideas about the twelve senses, composed of two groups, one of five and one of seven.

On 3 April 1917, again in Berlin, he explained the three principles of body, soul and spirit, as they were first given in *Theosophy* in 1904, through which true insight into the being of Christ first becomes possible.

These were new basic principles of natural science. They were deeds of knowledge by means of which natural scientists and doctors can pursue a different, non-materialistic course if they will only listen to it!

Rudolf Steiner presented his cognitive experience on three different levels—art, natural science and medicine, and social and political life—to those who would accept it.

In order to make the threefold idea effective for mankind, in order for it to have practical application in politics and the social life, its principles had to be proclaimed by some well-known, leading personality of the day. People were still living under the old hierarchical order and the new order should, by rights, have been introduced by responsible members of the old order.

Von Moltke, who had left the earth prematurely, would, through his karma, have been the most suitable person to announce the new teaching of body, soul and spirit, which had been repressed in the 9th century. But both he and his wife remain closely linked in soul with the will of the German folk-spirit and with Michael, the spirit of the age. Rudolf Steiner clearly stated this. But he was now obliged to find other personalities willing to proclaim the idea of the Threefold Commonwealth.

Therefore, at the end of May 1917 negotiations began in Berlin between Rudolf Steiner and Count Otto von Lerchenfeld of Bavaria, about introducing the idea of a threefold social order into politics. Count Lerchenfeld writes: 'As the Egg of Columbus—as a way out of chaos into peace, Rudolf Steiner confided his plan of the Threefold Social Organism to me.'[28]

Lerchenfeld himself, as a member of the cabinet, had the best connections to the leading personalities of that time. Thus, he was able to arrange an interview with Count Bernstorff, the German Ambassador in Washington, on 19 July 1917, at which Count Polzer was also present. But the discussion proved unsuccessful. Also interviews with Richard von Kühlmann, the then Foreign Secretary, were arranged.

Rudolf Steiner said to Kühlmann in July 1917, among other things: 'You have the choice: to listen to reason and take note of what the evolution of mankind proclaims to be a necessary action—for what these ideas embody is not any kind of programme, of which there are so many today, but it is something which has been drawn from the evolution of mankind and will most certainly be realised within the next 15, 20 or 25 years, but which especially has to be brought about in Central Europe. You have the choice today, either to see reason, to accept what common sense wishes to be realised, or else to advance towards revolutions and cataclysms.'[29]

Count Polzer brought a second memorandum to his brother in Vienna, the head of the Austrian Government, who had the best personal relationship to Emperor Charles. Rudolf Steiner advised Polzer's brother to try for the post of Foreign Minister and added: 'If the Austrian Emperor were to affiliate himself with this, then Bismark's saying might come true: "When the Austrian Emperor gets on his horse, all his subjects follow".'[30]

When Rudolf Steiner saw that Count Polzer could hardly bear the responsibility laid upon him, he comforted him by saying: 'One must also learn to stand back and watch things being destroyed. If what I have explained to you as a possible means of salvation does not come about, a series of catastrophes will happen. But that which cannot come about through reasonableness will happen nevertheless after great upheavals; for cosmic will requires it.'[31]

Emperor Charles, Count Czernin and Prime Minister Seidler had the ideas clearly explained to them. Count Polzer's brother had a talk of several hours with Emperor Charles, who ordered him to work out a memorandum. Polzer did so, sent it to the Emperor in a sealed envelope—but nothing came of it.

At the same time Rudolf Steiner had talks with Prince Max of Baden, the later Prime Minister of the German Empire. At these discussions Max von Baden fully recognised the possibility of salvation through the introduction of the Threefold Commonwealth and promised to bring it up in Berlin.

What a painful disappointment it must have been for Rudolf Steiner when, in his speech on 5 October 1918, Prince Max did not announce the Threefold Commonwealth but instead the 14 points of President Wilson and, on top of that, defended the terms of the Armistice of 3 October 1918.[32] Ludendorff had vigorously impressed upon the new Prime Minister that the plea for an armistice was absolutely necessary. After the Prime Minister's speech, however, he explained that he had made a mistake. For that he was dismissed by the Kaiser on 23 October 1918.

The Armistice between Germany and Russia was signed on 15 December 1917. Three months after the illusory treaty of Brest-Litovsk, Rudolf Steiner said about it on 24 November 1918:

> And this programme [the Threefold Commonwealth memorandum, which Kühlmann had in his pocket] would have been the only really effective programme if it had been brought up at Brest-Litovsk. Of course Brest-Litovsk would never have come about if there had been any understanding for such a programme as this. Things would have worked out quite differently then. For by that time I had worked it out as a

> guideline, not only for home politics, but for foreign policy too; home politics seemed to me to be superfluous when everyone was busy making armaments. What seemed necessary to me was a real impulse, which could have guided things in a different direction.[33]

In lectures during these months he said further:

> For reasons of human destiny, the war should end in 1917. Otherwise it will become something which will be completely inconceivable according to the old conception of what war is. For man would be completely excluded from control over events, and in place of events controlled by man, there would in future be a mechanical unfoldment of Ahrimanic effects.[34]

The attempt to lead politicians to recognise and proclaim the idea of the Threefold State as a final solution, went astray. Opposition was too strong. Mankind was led towards annihilation by Lenin in the East and by Wilson in the West. Rudolf Steiner repeatedly turned against these two opponents of the 'Representative of Man'.

Thus things proceeded towards chaos. The civilians and the soldiers in Germany were filled with a desire for the war to end. The soldiers wanted to go home. They despaired about this on-going tragedy, which was neither war nor peace. Revolt broke out. People wanted to achieve change through force and violence.

The leaders rejected the one real solution and did not make it public. They, who were formerly highly respected and even worshipped had failed and had to leave the stage of world events. Emperor Charles wrote to Kaiser Wilhelm on 4 November 1918 to say that he regretted having to conclude a separate truce. This truce had been dictated by Wilson and upset Austro-Hungarian unity. The monarchy disintegrated into a series of small states which were thenceforth dependent either on the East or on the West. Charles still hoped for a mutual and honourable peace treaty, but was obliged to abdicate on 11 November 1918 and leave Austria the same day. Kaiser Wilhelm had been obliged to abdicate two days earlier on account of pressure from America and in compliance with the revolution which flared up in Berlin. He went into exile in Holland on 10 November 1918.

At the same time Philipp Scheidemann proclaimed the Republic in Berlin, and after only six weeks in office, the Prime Minister, Max of Baden, handed over his post to Friedrich Ebert.

How was it this change came about so quickly?

Soon after the death of Helmuth von Moltke and after the trench warfare in the East and West had dragged on for many months, people began to rebel. There was talk of a second, 'vertical' folk-migration, from below upwards, which began in the East, continued in the South and, after October 1918, occurred in Germany too.

The fleets which lay at anchor in Kiel and Wilhelmshaven were to have set sail on 24 October, in order to engage with the English fleet in the Channel and thus indirectly relieve the Flemish army. But the sailors were afraid that a battle with the English fleet had been planned to take place in the Skagerrak so that they should perish in honourable warfare shortly before a dishonourable truce. They mutinied, and the ships which had left port returned to harbour. It developed into riots among the mutineers even in urban districts of Kiel. In Bavaria a mass movement started on 7 November on the 'Theresien meadows' in Munich, which spread to the whole of the city within a few hours. The revolution forced Ludwig III to abdicate on 13 November.

The fate of the German revolution, however, was concluded on 9 November in Berlin. One can discern, just in the events of this particular day, how it differed from other more guided revolutions: in Germany a crowd of people rose up with surging will force, but then waited for someone to come along to tell them what their goal should be. Because of this attitude the whole Threefold Commonwealth movement miscarried!'

Unnumbered crowds gathered on this 9 November in front of the Reichstag building and waited. Friends collected Scheidemann from the Reading Room and pressed him to speak to the crowd, for Karl Liebknecht wanted to make a proclamation from a balcony of the palace, as was done in Moscow, in the Soviet Republic. The Spartacists were already waiting with their red Hammer and Sickle flags.

Twice, then, on the same day, a Republic was announced in Berlin! From the speeches of Scheidemann and Liebknecht a comparison can be made which speaks for itself. Scheidemann: 'We want to form a workers' government to which all the socialist parties belong. We do not want to be a Soviet affiliation!' Liebknecht: 'We want a free social Republic serving world-revolution!'

Scheidemann preceded Liebknecht by a few hours and had already won the crowds over to his side. The Kaiser's government had abdicated and a provisional government under Ebert, Scheidemann, Noske and others had been set up on 10 November in collaboration with the army; and a random bloody revolution was thus avoided. But the new Weimar Republic, formed in Weimar in 1919, could only survive until Hitler's time.

But peace was at first short-lived. In upper Silesia terror surfaced and disturbances dominated the voting, as already described. On 13 March 1920 the Kapp-putsch took place in Berlin: Kapp announced a radical, right-wing government, which was to have been backed up by the army. It was a dangerous moment when these troops of the Right marched on Berlin. In order to avoid bloodshed the Government moved to Dresden. Then a general strike was announced and Kapp was forced to give up on 17 March, as he could not prevail against this passive resistance.

Directly after that, rebellions began in the Ruhr district. These had already started during the Kapp-putsch on 14 March and had spread like a lava-flow. The German army marched into the Ruhr district on 3 April and restored order. This fact was used by Clemenceau—'German forces have entered the demilitarised zone!'—as a reason to occupy Frankfurt and Darmstadt.

The Communists also tried to seize power in Saxony and Thuringia in October 1923, but they were defeated by the German army. One could mention quite a few rebellions between January 1919 and November 1923, in which new forms of government were announced. And on 8 November 1923, in the 'Bürgerbräukeller' in Munich, Hitler declared that the Weimar government (which he reviled) was 'deposed' and next day he undertook a march through Munich with General Ludendorff, which was scattered by police in front of the Feldherrn Hall.

Many small disturbances and a number of strikes kept Germany in continuous trembling disquiet, as also did the great railway strike of February 1922, which gripped the whole of Germany and hindered Rudolf Steiner and his wife, who were in Koberwitz at the time, from departing by train.

During these years there were also hundreds of political murders. Many people who appeared to be doing well in a political party were murdered by extremists of other parties. It would be informative and interesting to describe the circumstances, but that would lead too far in this context. Therefore I will only mention certain names. The following were murdered:

On 15 January 1919 Rosa Luxemburg and Karl Liebknecht in Berlin.
On 21 February 1919 Prime Minister Kurt Eisner in Munich, with whom Rudolf Steiner had talked about the Threefold Commonwealth and the question of war debts.
On 26 August 1921 Matthias Erzberger in the vicinity of Freudenstadt.
On 24 June 1922 Walther Rathenau in Berlin.

It was also, by the way, part of the plans of the 'National Radicals' to do away with the 'dangerous' Rudolf Steiner and his Threefold Commonwealth ideas. There had been a plan to assassinate him in Munich.

On top of all this disturbance and uncertainty came the devaluation of the currency, which affected every single person. No one trusted the German economy any longer and by 1918 the Mark had fallen to half its previous value. Before the war the U.S. Dollar was worth 4.20 DM.

In January 1920: 42 Marks;
in January 1921: 65 Marks;
in January 1922: 200 Marks;
in January 1923: 49,000 Marks.

In January 1923 a loaf of bread cost 400 Marks; in September it cost 8 million. Cities issued their own banknotes, and the Koberwitz firms introduced 'rye currency'.

In November 1923 the crisis was overcome by introducing a currency reform (Rentenmark).

During this unsafe time, everywhere seething with unrest, Rudolf Steiner poured all the powers at his disposal into his social impulses. He tried to realise his plans for the threefolding of the social organism in Stuttgart with the help of his friends Emil Molt, Hans Kühn, Karl Unger, Emil Leinhas, Wilhelm von Blume and others. He framed his 'Appeal to the German Nation and the Civilised World' and said on 21 April 1919: 'Today, when everything must begin with the general public, and when the October and November days of 1918 lie between us and the memorandum of 1917, the right path is to turn to the broad masses of the people.'[35]

The Appeal, which was launched on 27 January 1919, was actually intended for the whole world. It is an appeal to humanity as a whole. The social question which is there invoked has to be tackled afresh by each succeeding generation, in order for it to bear fruit for the following one. The appeal, which was signed by many hundreds of people in Europe, formed the basis for the work of those who were actively engaged in it. It was read by thousands. But even those who set their names to it did not take in what they were signing, and were not inspired to work to further it.

Rudolf Steiner began his work on the Threefold Social Organism in Zurich and hoped to reach out internationally to all humanity through it. Apart from that he hoped that at least in Switzerland he would find people who would acknowledge threefolding out of their sense for freedom. Unfortunately no one really understood his efforts; and on Easter Saturday 1919 he said:

This is one of the saddest things of the present day. This brutal response arising from the most terrible lack of understanding for what would be of benefit to mankind: 'We cannot understand it, it is abstract...'—or something of the kind. It is just those people who willingly accept all the censure or censures of various countries, and who repeat in parrot fashion all the sayings which come from above, be they ever so foolish, who are unable to understand something which is directed towards their free minds, their free souls! But today we stand at a point in time at which only that is decisive which is taken up out of free understanding; only that is important which a person is not forced to acknowledge, but which he desires to understand out of the depths of his heart.[36]

Rudolf Steiner travelled to Stuttgart on Easter Sunday, 20 April 1919, and there took upon himself the superhuman task of creating a Workers' Movement. The book, *Towards Social Renewal* (*Kernpunkte der sozialen Frage* in the original German) appeared on 28 April 1919, was translated into almost all European languages in thousands of copies together with the 'Appeal', and was read in every country of the world. There followed other appeals and lectures in the great industrial factories in Württemberg and Baden, especially in the Daimler and Bosch factories, together with discussions, conferences and private conversations.

The founding of the Threefold Commonwealth Association, the forming of business and cultural advice centres, the founding of the *Kommenden Tag* in Stuttgart and the *Futurum* in Switzerland occupied Rudolf Steiner day and night. On 28 May 1919 he wrote to Eliza von Moltke:

I stand here in a virtual cross-fire. The more what I want is achieved, the heavier is the onslaught of the opponents. Every evening there are lectures or negotiations. The worst is still to come, for we are only at the beginning and people continue their soul slumbers. In middle-class society—and the leaders of the Social Democrats also belong to this class—social sleep everywhere prevails. Out of this non-social condition, sociality must be created. The problems are compounded by people's lack of understanding. The most important of my intentions are simply not taken in by them. It is just as though they cannot comprehend anything to which they have not been accustomed for the last 30 years. Sclerotic brains, paralysed etheric bodies, vacant astral bodies, completely dull egos. That is the signature of modern mankind.'[37]

Rudolf Steiner could see that the proletariat took in what he said, but the Trade Unions did not. The industrialists certainly opposed him in every way they could, so that by 21 January 1919, as a last resort, he implemented his already planned idea of founding a 'free school'. On 2 November 1919 he drew his conclusion about this period: 'Civilised man of the present day will either have to acknowledge an independent spiritual life, or our present civilisation will meet its destruction; then a future will have to evolve out of Asiatic cultures.'[38]

Rudolf Steiner, with the help of his friends Molt, Stockmeyer and Hahn, succeeded in carrying out preparations for the founding of the Free Waldorf School as a comprehensive school. They found their site in a café on the Uhlandshöhe, the destination of trips by many of the inhabitants of Stuttgart. The opening ceremony took place on 7 September in the City Gardens, with the participation of the population.

19 September was the first day of school for 256 children, the parents of most of whom were workers in the Waldorf-Astoria cigarette factory. Rudolf Steiner was the School Director, Herr and Frau Molt were the 'school parents'. In his speech in the City Gardens Rudolf Steiner spoke about the central task of this school: '...enlivening of science, enlivening of religion, enlivening of art'. He closed with the words: 'What is experienced today, especially in Central Europe, as the greatest need—is only the beginning of what may be experienced as even greater need.'[39]

One might very well say that Rudolf Steiner's intention in founding the Waldorf School, was to plant a seed so that, in times to come, when chaos and catastrophes would once more, inevitably, give rise to the idea of a threefold social order, there would be people who might possibly be able to comprehend it.

There is one more thing of great significance to be mentioned, about which Rudolf Steiner tried by all means to establish the truth and was hindered by the personal interests of others—the question of war blame.

If one takes the trouble to study the documents relating to negotiations between Central Europe and the Allies, one is overcome by the amount of hate and mendacity, the ill will and false conclusions, the number of senseless, so-called punitive measures and acts which ruined the future of Central Europe—all because Germany was held to be the only country responsible for starting the war.

Everyone should at some time read the documents about these events! Then it would be quite clear that Rudolf Steiner had told the truth about these events and their background.

The enemies of Central Europe wanted to destroy it, in full consciousness—although this was only achieved after the Second World War, by the division of Germany. But every thinking human being can understand that such vengeful acts, at odds with the intentions of the folk-spirit, would lead to new wars and catastrophes.

Already on 12 December 1916, Germany and its allies made an offer of peace to enemy governments through the Swiss Ambassador and the American Chargé d'Affaires.[40]

This offer was rejected on the grounds that Germany was the only one to blame for the war and therefore had no right to offer peace.[41] Germany also offered peace later, as for instance the peace resolution of the Reichstag on 19 July 1917.[42] The Stockholm Memorandum of 12 July 1917 tried to bring peace to the world via all the Social Democratic parties.

Even the peace initiative of the Pope on 1 August 1917 was again rejected by the Allies on the grounds that Germany had started the war in order to take over world supremacy. On 28 September 1918 Ludendorff then pushed for an immediate armistice after what, as he said, was the defection of Bulgaria, the forthcoming separate armistice of Austro-Hungary and the merely burdensome alliance with Turkey.

Prince Max of Baden acceded to this and asked President Wilson for a truce on 3 October. In his speech on 5 October, two days later—as already mentioned—instead of announcing the Threefold Commonwealth, he laid the foundation for the Versailles Treaty.

Everyone thought that the 14 points announced by Wilson on 8 January 1918 were an adequate basis on which to guarantee world peace. But when one now reads the speeches and proclamations of Wilson of 8 January, 11 February, 4 July and 27 September 1918,[43] they strike one as being the Utopian dream of an inexperienced simpleton. They contain cleverly thought-out illusions without any sense for reality, similar to the Brest-Litovsk treaty. Walther Rathenau described Armistice Day as the 'dark day' of German history.[44]

Others, too, were indignant, as for example Scheidemann. Ludendorff was certainly dismissed, but the Armistice took place nevertheless. Ludendorff's report about proceedings connected with the peace offer of 31 October, gives a great deal of enlightenment.[45] There he wrote that it was quite clear to him that Wilson wanted to destroy Germany.

On the basis of these events connected with the Armistice it was made possible for the Allies to force the Versailles Treaty onto the Germans. During the negotiations Wilson was quite besotted with his abstract 14 points and by his intention to destroy Central Europe.

Lloyd George, inspired by the thought that England was the only country with world tasks to fulfil, was worried about Germany's possible economic rivalry. Clemenceau was filled with hate and egoistic-nationalistic impulses against his 'fear-inspiring' neighbour. He searched for a means of justifying the destruction of Central Europe—for there could be no talk of negotiation! And the only reason which all the world accepted without proof was the assertion that Germany bore sole responsibility for the war, and therefore had to take all the blame for its horrors and the streams of blood which it had shed. Germany had not only to be punished, but also humiliated in such a way that during the decades to come it would be dependent on the control of the allied powers and unable to offer them any economic competition. Clemenceau's note to the German Foreign Secretary, Count Brockdorff-Rantzau, on 16 June 1919, expressed this destructive desire quite clearly.[46]

And also the discussions between Professor Delbrück and Colonel Conger of Wilson's staff reveal quite openly the intention of the Allies to destroy Germany and to acknowledge the blame for the war as its justification, even though Wilson endeavoured later to reduce the severity of the conditions a little.[47]

People of many countries were indignant about the Versailles Treaty. The German government under Scheidemann, and with it the Foreign Secretary, Count Brockdorff, resigned on 20 June 1919. The parties spoke up against the signing of the Treaty. But, finally, the compulsion of the food-blockade, and the threatened invasion of Germany by the Allies induced the new head of government, Gustav Bauer, to agree to the signing after the German National Congress had also agreed on 22 June.[48] So this universally disastrous treaty was signed on 28 July 1919 by the Foreign Secretary Hermann Müller and the Colonial Minister Dr Johannes Bell (who only lasted till November 1919).

The results of these events, of such hindrance to the will of the German folk-spirit—National Socialism, the Second World War, the division of Europe—do not need to be specifically mentioned. We are all suffering from them today, and it would be good to remind ourselves of the failure of political leadership at that time, because the dealings of today have consequences for the future.

How differently would events have worked out if Germany had had the courage to refuse to sign the treaty—or if the suggestions of Rudolf Steiner had been accepted!

In countless lectures and discussions Rudolf Steiner tried to make the events between 1914 and 1919 clear and to explain their background and reveal the truth. He tried to prove what had really happened at the start of the war in 1914 by publishing von Moltke's memoirs. And because this section of history is of special interest and has been experienced by many of us, it seems to me to be necessary to briefly mention aspects of it, through which the importance of the Koberwitz impulse can be recognized.

Rudolf Steiner pointed out in his lecture on 14 October that direct effects and causes, especially for the periods between 1912 and 1914, have to be looked for in events taking place in the spiritual world. Which events is he referring to? In 1841 the battle of Michael against the dragon began. It ended in 1879 with Michael's victory and the fall of the Ahrimanic Spirits to the earth—into human beings. The year 1879 is also a historic beginning in another respect, for the rule of Michael as an Archai began. 33 years later, in 1912, the Anthroposophical Society was founded in Cologne. Rudolf Steiner often said that we should observe the mirroring of events on earth. For instance, an especially vigorous battle took place in 1844/45 in the spiritual world, that is about 35 years before 1879; and 'when such a cycle has been completed'—that is 35 years after 1879—a special event happens on earth: the start of the war in 1914.

Michael's battle began 38 years before 1879; exactly the same number of years after 1879 the decisive turning-point in human evolution came about in 1917. On 6 February 1919 Rudolf Steiner named another year in the middle of the 19th century as being particularly important. It is the nodal point at which the modern materialistic way of looking at the world began: 1859, the anniversary year of Schiller's birth. Schiller, whom Rudolf Steiner described as the genius of the German Nation! But what came to expression in the speeches of the Schiller-year showed that idealism had become mere phrase. Progress now consisted in getting to know what is inert, dead.

Karl Marx published the first materialistic history book in 1859: *Towards a Critique of Political Economy*. Kirchhoff and Bunsen tried to prove the materiality of the stars by means of spectral analysis. Darwin started to teach the 'Origin of Species'. Since that time no further evolution of humanity is taken into account as a real possibility.

Long ago the idea of the threefold nature of man had been driven out of European consciousness by the Ecumenical Council of 869.

Rudolf Steiner's own principles did not allow him then to speak openly about the fact that the war had been intentional on the part of occult circles in East and West.

Today this can be proved through documentary evidence. Historians now describe things which Rudolf Steiner had already said in 1917/18:[49] how, for instance, the society of the 'Black Hand' wanted to divide Southern Europe from Austro-Hungary to bring it under Russian 'protection', but was only properly successful after the Second World War. Every method was used to achieve that end: slander, murder, and especially war!

The most important documents, however, which can prove the true events at the outbreak of war, are the memoirs of the then Chief of General Staff, Helmuth von Moltke. The Prime Minister, Eisner, possessed other documents—for instance, one which stated that a large consignment of munitions had been cancelled shortly before the outbreak of war[50]—so the head of the War Department could not have reckoned with the possibility of war—but Eisner was shot before he was able to publish this.

On 27 January 1919 Rudolf Steiner conceived a plan, together with those who were gathered in Stuttgart, to publish Moltke's 'Memoirs' along with the 'Appeal', as a necessary explanation of the causes of the war, in view of the impending peace negotiations. Rudolf Steiner had wanted these memoirs and the history of the outbreak of war to have been published in Stuttgart and Switzerland in 1916 already. This was in order to inform the international public that all the blame did *not* rest with Germany, which was the lie used by the Allies when they rejected Germany's peace offers.

The correspondence between Rudolf Steiner and Frau Eliza von Moltke in May/June 1919 showed how much Rudolf Steiner was concerned to have the book brought out before the end of the peace negotiations. On 3 May he wrote:

> By making these notes known, a healthy start can be given to a possible peace treaty, both outwardly and inwardly—the present Armistice in Versailles is, of course, an impossibility, however things may turn out. And without a possible peace

Germany will not be able to survive, even under the best circumstances, but will only continue to fall into ruin.[51]

The brochure with the Moltke memoirs and an article by Rudolf Steiner on the war-blame question, was ready for distribution on 27 May 1919. The Association for the Threefold Social Organism added an appeal to it with a print-run of 50,000 copies with the title: 'To the German People and the German Government'. The appeal included three demands for ascertaining the truth about blame for the war. The brochures and appeal were to be laid on the negotiating table in Versailles by the Secretary of State, Dr Schall. The Allies would then have been unable to maintain the lie about the sole guilt of the war resting on Germany; and the Versailles Treaty would have had no foundation in fact! But adversarial forces used all means to hinder this intention.

'Anthroposophists' got hold of copies of the brochure a day before it was due to be distributed and so it was made known prematurely to the public and to the Stuttgart Parliament, before the 50,000 copies could take them unawares.

Rudolf Steiner wrote to Frau von Moltke on 28 May 1919 when he sent her the first copy: 'Just as the first brochures reach me I hear to my horror that "anthroposophists" had collected copies yesterday. It is terrible that among anthroposophists no kind of order can be established.'[52]

On the very next day, 29 May, General von Dommes came to visit Frau von Moltke. Then, on 1 June, he visited Rudolf Steiner in Stuttgart, four days after the latter had received the first copy of the memoirs. He wished to assure Rudolf Steiner on oath that the Moltke memoirs contained three essential errors. Hans Kühn reported that the eldest member of the von Moltke family had thereupon asked for the memoirs to be withheld, as the honour of the von Moltkes 'could not allow the Kaiser to be shown in such a bad light'.

Rudolf Steiner wrote to Frau von Moltke on 6 August 1919:

> What the Chargé d'Affaires Counsellor von Moltke [the eldest member of the family mentioned above] did on behalf

of the Foreign Office would never have led me to comply. I had to agree, though, on account of the intervention of von Dommes. He came to me and explained that he could prove that the memoirs contained three points which were not in accordance with the facts ... there is no doubt about it, that the soul [of Helmuth von Moltke—Chief of the General Staff] still believes in the truth of the report.[53]

The adversarial powers had succeeded in preventing Rudolf Steiner's words from being accepted in Versailles, and that is the reasons why the misfortunes of Germany will endure for decades to come.

At this time Rudolf Steiner constantly referred to Wilson, for instance on 20 July 1919: Wilson's 14 points, he said, are a Utopia which has echoes of the Tower of Babel—they are supposed to have solved the task of uniting the nations but will only serve to drive nations further apart. The chaos which arises as a result will exceed that caused by the Tower of Babel.

Only in 1924 did Rudolf Steiner speak about the former incarnation of Wilson, in which he had been connected with the Arabian stream in the 7th Century—as Muavija, one of the earliest followers of Mohammed, who therefore had to oppose Central European spirituality. But hardly anyone paid any attention to what he had to say, let alone acted in response. In many lectures and conversations, Rudolf Steiner described Wilson's intentions and always tried to counteract the lie about the blame for the war. But the Versailles Treaty was forced through and the submission of German people to the domination of the western powers began.

In his lecture on 20 July 1919 Rudolf Steiner made clear what had taken place. In conclusion he defined three aspects of modern society:

1. The attainment of world domination by Anglo-American powers.
2. The endeavour to unite nations, which at present takes quite abstract forms (League of Nations).

3. The endeavour to imbue world affairs with social understanding.

There were tremendous obstacles, he said, working against these three endeavours. Against what was striven for by the Anglo-American world, issuing from England as world domination, was the spirituality of Ancient India. That would, he said, lead to great differences: on the one hand western power-seeking, on the other hand the search for world-principles through the Yoga path. This difference will lead to the greatest spiritual battle to be fought in world history. The first task of a true follower of spiritual science is to discern accurately the two poles in current events.

The action taken against the war-blame lie was thwarted, as were Rudolf Steiner's efforts to establish the Threefold Social Organism, considered by him to be a failure. Already in November 1919 he ceased to work further on it.

On New Year's Eve 1922/23 the Goetheanum burned down, so that those dependent on outer observation were henceforth unable to experience the spirit of Central Europe through it.

Outwardly there was no prospect of any redemption. Then Rudolf Steiner decided to make the greatest personal sacrifice that any initiate had ever made. He resolved to hold the Christmas Gathering at which, with pain, he took over Presidency of the Anthroposophical Society so that he could then talk concretely about questions of karma.

He now imparted a knowledge of the threefold processes in the esoteric realm, as cosmic-human process, and laid the tripartite Foundation Stone in the hearts of mankind in a meditation which left man free.

Instead of accepting these spiritual truths in devotion and reverence, however, the members started to quarrel a few days after Rudolf Steiner's death, as to who had which rights and who was practising the right methods, and many other things. But spirit-beings and the dead flee squabbles as we do fire!

In this way the being of Anthroposophia withdrew from

the Anthroposophical Society. Man did not take up the words of truth.

Countess Johanna Keyserlingk had such a profound experience during the days of Rudolf Steiner's death, that it should be added here to the foregoing. This also needs to be done to put right what was quite incorrectly published by others, without the knowledge of the present editor.

Rudolf Steiner was able to instruct his pupils through his spirit body, as this is described in the Mystery Dramas. Already some instructions of this kind had taken place, so that those which are described here should not cause a great deal of excitement, but can be accepted in a quiet state of mind:

> It was on the morning of Rudolf Steiner's cremation, which I had not been able to attend. The earthly body of the exalted teacher still stood nearby, laid out in state in the *Schreinerei* [workshop room], when the aura of the beloved teacher appeared to me. From the latter issued the message that I should write something. I took paper and pencil, and from his presence came the following words. Often I could not keep up with it, then there would be a pause and a wait until I had caught up, just as Rudolf Steiner used to do before when he dictated to me.
>
> 'My mission has come to an end.
>
> What I could impart to the maturity in man I have given him.
>
> I go away, for I found no ears which could perceive the spirit word behind the word.
>
> I go away, for I found no eyes which could perceive the spirit-pictures behind the earthly pictures.
>
> I go away, because I found no human beings who could bring my work to fulfilment.
>
> The Mysteries remain veiled until I return.
>
> I shall return and reveal the Mysteries when I have succeeded in building an altar, a place of worship in the spiritual world for the souls of men. Then I shall return. I shall then continue to open up the Mysteries.
>
> They are to blame for my death who have impeded the culture of the heart.

> If people had delved deeply with their hearts into the depths, they would have found the strength to fulfil the tasks of the age satisfactorily.'

When he spoke to the young people at Koberwitz, too, Rudolf Steiner talked of the altar beneath the earth, and in the last scene of the Mystery Dramas is mentioned the Light which burns Ahriman when it is kindled in the pupils of Benedictus. In another place mention is made of altars in Michael Sanctuaries or Schools, on which flames must be kindled.

Emil Bock wrote the following about this period:

> When Rudolf Steiner came to Stuttgart in May 1913 he spoke for the first time on German soil about the secrets of the Archangel-periods and about the epoch-boundary of 1879, in the lectures entitled 'The Michael Impulse and the Mystery of Golgotha'. The cloud of the First World War had already started to darken the sky [...] Rudolf Steiner was in England. He spoke prophetically about the Apocalypse, which was to become a matter of first-hand experience through the war, and he uttered the Michael secret: Michael is the one who provides the apocalyptic Light for what is approaching. And in the Light of Michael, at the same time, the starting point of 'real Anthroposophy' became visible in the teaching about the senses. In 1916, as new waves of destiny approached, Rudolf Steiner was so closely connected with the Archangel of the Age, as also with the German folk-genius, that he was able, before the storm broke, to utter prophetically what was seeking entrance into mankind through the spirit. That was coloured by the meeting of Rudolf Steiner with Helmuth von Moltke. The sphere of Michael and of the folk-genius revealed itself and finally enabled the step to 'Anthroposophy' to be taken.[54]

Bock goes on to describe how, after his obituary speech for Moltke in the Berlin Branch, Rudolf Steiner spoke publicly for the first time, on 20 June 1916, about the 12 senses and thus introduced a new basic principle of natural science.

> That is nothing less than 'enchristened natural science', to know nature in such a way that this knowledge, in its very

language, corresponds with the Etheric Christ. It becomes quite clear that 'Anthroposophy', in this particular sense, leads to the conquest of untruthfulness or even of unchristianness, which have taken hold of human thinking and knowledge.[55]

And once again I shall quote Emil Bock, because I am unable to express what I want to say better than this:

> But wherein lies erroneous thinking? It lies in the attitude of natural science which wants to observe only the material, external side of existence. If one wants to get to the root of all disasters, including the catastrophes of war, then one must not only take the step from natural science to spiritual science, but one must, above all things, take the step from spiritual science to the new natural science, because only through a new kind of natural-scientific thinking, through a spirit-enlightened thinking, even about the human organism, can the starting point be created for social action.
>
> When Rudolf Steiner returned to Berlin after this half year he took up the third element. There are three elements: the 12 senses, representing the human physical body; the 7 life-organs, which serve the etheric body; and then there is also the threefold human organism as a whole, which in its physical organisation is the bearer of the soul in thinking, feeling and willing.[56]

Rudolf Steiner, who is inwardly connected to Michael and to the German folk-spirit, battled for the free development of the human ego. In 1899 Kali-Yuga came to an end. He stood in the light which so blinds mankind that it can only see the darkness. He saw that the etheric earth-aura had become pervious to the sun-forces, to the Etheric Christ, since 1909, and he founded the Anthroposophical Society in 1912 in order to bring the light of thinking into mankind's consciousness, to penetrate the darkness of its ego. But mankind rejected it; and the adversarial will to destroy the spirit of Central Europe asserted itself.

Helmuth and Eliza von Moltke are closely connected with the Light of Michael which, as the third principle, the spirit, has to flow into mankind again, for in a previous incarnation they had to point Ahriman in the direction of knowledge without spirit.

In association with the von Moltkes and those who have joined with them in the School of Michael, Rudolf Steiner described the threefoldness of world, earth and man in the seven liberal arts and made use of the categories of Aristotle for that purpose. But mankind rejected these spirit-deeds. It hindered the threefolding of the social organism. It would not accept the teaching about the three-times-four senses. Rudolf Steiner perceived the threatening abyss and annihilation and gave a new message to mankind at the Christmas Gathering.

His opponents tried to poison him, but he overcame it.

They wanted to murder him in Munich. He stood on the stage, the electric current was shut off: Rudolf Steiner was the only one in the hall who was lit by a candle. He calmly went on speaking and no one dared to shoot him!

He was personally protected against outward powers, but the failure of the heart-forces of those close to him made him ill. People did not accept his words, although Michaelic intelligence had flowed for centuries upon mankind.

In the last year of his life, 1924, Rudolf Steiner called down heaven and its wisdom onto the earth, but it was not understood by man. In a conversation with him Countess Johanna Keyserlingk said:

'Humanity is very much mistaken', whereupon he answered:

'Yes, one can certainly say so.'

'But what will happen at the end of the century?' Thereupon he answered in a despairing tone of voice:

'I cannot tell—but Ahriman is also a part of Christ.'

I learned about this conversation soon afterwards and his answer has accompanied me throughout life.

Rudolf Steiner created three basic ways—apart from the various training courses—to enable one to take up anthroposophy into one's heart and will in a quite personal way.

The most personal and intimate way was through the portrayal of the representative of mankind between Lucifer and Ahriman in a statue which he carved in wood out of the strength of his active will. Everyone who inwardly unites himself with the threefoldness in and around this representative of mankind can recreate it within himself, and form and conduct his life through it.

The other way was by teaching about the threefold process in four, seven and twelve factors in nature and in man. Every single person has to attain his own, independent, conscious relationship to it. Without experiencing these three processes no doctor who wishes to be an anthroposophist can make a living diagnosis, let alone work out a therapeutic plan.

The third way is the Koberwitz impulse, which encompasses the threefold principle both on and in the earth. Every farmer and gardener has to acquire a personal connection to the Agricultural Course and keep it alive within himself. Otherwise he will not be able to organise himself with his community of man, animals, plants and the soil and its environs in the right way. He must experience what is organically alive, otherwise he will not be able to understand a living kind of thinking. If he thinks materialistic thoughts, or even of financial considerations, he will have to make compromises which repel any kind of spiritual or elemental world, and then he will declare that it 'doesn't work'.

But anyone who thinks thus has invoked the very misfortune which we now experience as the consequence of unreal thinking of the past. Everyone who believes that he has to comprehend the threefolding principle through a scientific-materialistic kind of thinking will create further disasters for the future.

The Koberwitz impulse concerns every single person. Every farmer is personally responsible for what he does—every one must become an individual ego-personality if he would apply the directions of Rudolf Steiner. If modern farmers of this kind would then unite together into communities in which each member can remain independent, then the social basis for the life of the Koberwitz impulse will have been created, for it is through food that we are provided, first of all, with the possibility of spiritual and social thinking.

Also the pollution of the environment, and the poisoning

of foodstuffs, which is generally recognised today with all the problems and dangers it entails, began in about 1912. Not only did the artificial fertiliser industry grow enormously, but also the use of petrol, and pollution through machines and radium began their triumphal advance on earth. Food will become so impaired that human beings are no longer able to think spiritually. Their brains will become salted and hardened. The instrument of thinking will be destroyed.

The Koberwitz impulse provides the basis for a proper nourishment, through which transubstantiation will still be possible. For making medicines one also needs to have unpolluted animals and plants.

It was not without reason that Rudolf Steiner spoke about 'islands' which would be necessary for the cultivation of German spiritual life. Spiritual consciousness and inner peace are being upset on all sides by food catastrophes and disturbances in the social structure of man's environment. Cities are increasingly becoming places where Ahriman can work unhindered. Only in monastic isolation, where the surrounding land has 'sunk' to an agrarian condition and where factory chimneys no longer pollute the environment, will Michaelically inspired people find the courage to create 'islands', where people can put the threefolding process into practice in all four realms of nature—people in whom the knowledge of the threefold nature of body, soul and spirit, lost since the Ecumenical Council of 869, is again living.

The two von Moltkes have been karmically connected to these impulses since the 4th and 9th centuries and endeavour to revive the knowledge of the trinity of body, soul and spirit in Central Europe with all the powers of their souls. The German folk-spirit is alive in these two personalities and therefore they co-operated with Rudolf Steiner when he brought these impulses to the consciousness of mankind.

Those who were spiritually awake during the years 1912 to 1924 called out:

> 'Veni creator spiritus!'
> Come, Creator of the Holy Spirit!

For the time is at hand! The spirit no longer comes of its own accord, as in former times, but the Creator of the Holy Spirit has to open the gates for it.

Did not Rudolf Steiner have to use his last Whitsuntide to reveal this Pentecostal message because it concerns the continuance of the earth and of earthly nourishment? Even today he still places it in the realm of each person's quite individual experience. Every human ego can accept these words in his heart, or reject them.

So this last Whitsuntide message of Rudolf Steiner concerns everyone who is of good will and who wishes to see in its words an ultimate redemption.

Appendix 1: Requiem by Rudolf Meyer (shortened)

On New Year's Eve (1928), having been summoned to the funeral ceremony, I stood beside the bier of Count [Carl] Keyserlingk. There arose before me in that hour the picture of another who had passed away, with whom I had just as strong a connection in anthroposophical activity as with our revered dead friend. It was the memory of our friend Louis Werbeck to whose bier I was led at the beginning of the year. How linked together are these two events of 1928. For these two men, who may be looked upon as pillars of the life of our Society were both called away in the midst of their activity; and of both we may say that their hearts were broken because they beat so strongly for all that is truly human, glowing for great earthly ideals, yet having to endure pain because of the resistance to these aims of the spirit of our age.

For Count Keyserlingk, though with more unexpected suddenness, also died of a heart attack. It was the night of 28–29 December, just as he had left Schloss Sasterhausen for the journey to Dornach, there to attend the meeting of the General Secretaries. He had set out full of fruitful plans for the furtherance of our work, deeply concerned about the welfare of the Anthroposophical Society. With his soul set towards Dornach he entered the spiritual world at midnight on one of the thirteen Holy Nights. It is not a matter of indifference for the destiny of the soul after death in what hour the spiritual world is entered. To awaken in the light of the Holy Nights is a grace bestowed by destiny; it is the time when the earth rays most brightly into the cosmos, when it longs to bear forth the Christ Light into the cosmic cold—the message of the earth! And this karma is in harmony with the being of Count Keyserlingk, for he was, if one is to characterise him, a man of Christmas. Those who were privileged to keep the Christmas Festival with him, must ever preserve in memory the picture of the opening of the doors of the great hall at Koberwitz—the same which has become so dear in the memory of those who listened there to Rudolf Steiner's Agricultural Course—where Count Keyserlingk stood alone between the tables heaped with carefully chosen gifts, the lighted Christmas tree behind him. Holding open in his hand the Gospel of St Luke, he would read to the many assembled guests from the household and the estate the Angel's message to the simple Shepherds. For the wisdom of anthroposophy so worked in his personality that he knew that the spiritual light now breaking into the darkness of the earth could indeed be received by those who are aware of the holiness of earthly nature.

When towards the end of the Great War, during which he had taken part in many great undertakings at the Front, he came into touch with anthroposophy, he immediately recognised its reconstructive power for the shattered regions of Central Europe. From that moment his whole will was set on bringing to their aid the wholly practical and necessary bases of anthroposophical methods.

The highest point of his rich life was reached when, at Whitsuntide 1924, Rudolf Steiner and more than a hundred guests could assemble day after day in his house, there to receive from anthroposophy the fundamental principles of a renewal of agriculture. They were festival days—the daily bread seemed re-sanctified. All could feel a newly dedicated relationship between the human being and the tilled soil, the mighty, nourishing mother. It was Count Keyserlingk's deepest longing to bring down the Christ Light into the most earthly realm, into the soil which man had ravished, into that which most needed redemption.

Through being associated with an estate, managed on a large scale, but also in accordance with the most modern and materialistic principles, Count Keyserlingk had realised the urgent need for a complete renewal of cultivation methods. It was because of his constant labours for the fulfilment of the impulses given in recent years by Rudolf Steiner, that our teacher gave him the name of the 'Iron Count'. But the iron which worked in his activities was the iron which rays out from man's heart as a living, strengthening power in initiative and actions.

The last picture that impressed itself on my soul was the sword resting on the coffin quite covered with red roses—only the cross-shaped hilt was visible; the blade was hidden beneath the blossoms.

As I was leaving the mansion, after the ashes had been put in their resting place, I was met with the shocking news of the death of Dr Unger. I drove immediately afterwards through a snow-covered country, glittering under the rising stars. It was the last of the thirteen Holy Nights before Epiphany. Dr Unger, too, had passed during this holy time towards the Star. His soul, weaving in pure thought-powers, seemed to my inner vision to merge into the starry cosmos. The pure world of ideas streamed towards him with a wonderful spiritual warmth. The need of his soul was to make the revelations of spiritual research transparent for man's thinking; human understanding and divine wisdom were to be reconciled to one another again.

Through its final destiny his spirit, so suddenly and mightily awakened to spiritual vision (he had been shot), seemed to me to be immersed in reading—better said: to be borne up into—the widespread script of the firmament. The earthly path of this seeker of the spirit now stood before my soul as that of the Wise Kings from the East, who discovered the Christ Star by deciphering the signs of the heavens.

But assuredly, there above, that other soul already kept watch in the open hall, for I must needs recall the last words spoken by Count Keyserlingk before he started his journey: 'The first thing I must do when I reach Dornach, is speak with Dr Unger'.

As two representatives of our movement, both drawing near to the light of the new revelation of the spirit, in fully human though quite different ways, these now come to life before my soul in their deathless being. Thus they inclined to the Christ Sun, of which at the great Christmas Foundation Meeting Rudolf Steiner spoke these words:

> Light that illumines
> The wise heads of Kings,
> Light that gives warmth
> To simple shepherd-hearts!

(From *What is happening in the Anthroposophical Society*, 27 January, 1929)

Appendix 2: Letter from Rudolf Steiner to Count Keyserlingk, 1919

Esteemed Count Keyserlingk!

Since talking with the Countess I find it necessary to add the following to what we talked about on your last visit. In the case of things which happen in the physical world, which are in agreement with our own will, we can say it is the intention of divine powers that such a thing has come about. This is not the case, however, when things do not work out according to our intentions. In the latter case Ahrimanic powers, working through human beings, become involved. And when this happens it often means that we have to work together with the good divine forces in order to combat the forces of hindrance, which not only oppose us, but also work in opposition to the divine will. The good forces rely upon our active participation.

In the case we spoke about on Sunday in Berlin, something was not immediately clear to me owing to the way you put your question, dear Count. Now the Countess has told me that the reason for your allowing yourself to be diverted from your original intention, is that you think that this adversity is the outcome of divine will. After further contemplation I feel that I have to say to you what I have written above.

I now have to tell you that the course of events—if what the Countess says about your attitude towards them is true—must not be allowed to miscarry because you think that divine will is opposed to your undertaking.

In such matters we often overlook the fact that, when divine will is active in the physical world, it unites with the human will in order to fight with the human being against Ahriman. Then it is right to comply with divine will. And we can rely in this case upon what our practical common sense tells us is the right course to take.

As mentioned above, this was not discussed in Berlin on Sunday, because the pertinent question was not raised. If that had happened, it would certainly have been my duty to say what is here expressed.

With respectful greetings

Rudolf Steiner

1. Koberwitz. Between the small first entrance, in front of which Rudolf Steiner greeted the participants of the course, and the main entrance beneath the balcony, was the great dining-room in which the lectures were held.

2. Inside the hall, a sketch. (Courtesy of David Clement.)

3. Countess Johanna von Keyserlingk, née von Skene, 1924/25—44 years old.

4. Count Carl von Keyserlingk at about the time of the Agricultural Course, at the age of 55.

5. Eliza von Moltke (died 30 May 1932 in Amberg/Starnberg Lake). (Courtesy of her granddaughter Frau Via Burgheiser, Erding.)

6. Count Carl von Keyserlingk. (Courtesy of David Clement.)

Notes on the sources and literature of Part III

1. Emil Bock: *Rudolf Steiner.* 2nd edition. Stuttgart 1967 p. 292 et seq.
   Helmuth von Moltke: *Erinnerungen, Briefe, Dokumente 1877–1916.* ('Recollections, letters, documents') Stuttgart 1922. (See *Light for the New Millennium*, ed. T.H. Meyer, Rudolf Steiner Press 1997).
2. Walter Goetz: 'Kaiser Wilhelm II und die deutsche Geschichtsschreibung' ('Kaiser Wilhelm II and the writing of German History'). In *Historische Zeitschrift* Vol.179, 1955, No. 1 p. 37. Helmuth von Moltke. As above, p. 19.
3. *Weltgeschichte in Bildern* ('World History in Pictures') Vol.22: 'The Epoch of Imperialism. The First World War.' Edition Rencontre, Lausanne and Librairie Hachette, Paris 1970, p. 109.
4. Jürgen von Grone: *Wie es zur Marnesschlacht 1914 kam* ('How the Marne Battle 1914 came about'). Author's own publication. Stuttgart 1971. p. 14.
5. Walter Goetz. As above.
6. Helmuth von Moltke. As above. p. 387.
7. ibid.
8. ibid (text of the Kaiser's telegram).
9. ibid p. 388.
10. *Dreigliederung des sozialen Organismus* ('The Threefold Social Organism'). Periodical. Year 3, No. 15 Oct., 1921.
11. ibid. No. 16, October 1921.
12. *Ursachen und Folgen. Vom deutschen Zusammenbruch...* ('Causes and effects. From the German collapse 1918 and 1945 to the New State regulations of the present. A collection of deeds and documents relating to the history of that time'). Published and edited by Dr Herbert Michaelis, Dr Ernst Schraepler and Dr Günter Scheel. Document publishers Dr Herbert Wendler and Co, Berlin 1958. Vol. II, No. 271. Quoted in future notes as: 'Documents'.
13. ibid. No. 305a, 305b.
14. David Shub: *Lenin—eine Biographie*, Wiesbaden 1957, p. 351.

15. ibid.
16. ibid. p. 445.
17. *Weltgeschichte in Bildern* ('World History in Pictures'). Vol. 23. 'The Treaty of Versailles and the post-war period'.
18. Documents. Vol.III, No. 689.
19. ibid. Vol.II, No. 399 a/XIII.
20. *Nachrichten der Rudolf Steiner Nachlassverwaltung...* ('News from the Administration of Rudolf Steiner's Estate with publication from the Archive') Dornach 1966. Nos. 27/28, 1969, p. 43. Quoted in future notes as 'News-sheet'.
21. Count Adalbert Keyserlingk: *Vergessene Kulturen in Monte Gargano* ('Forgotten Cultures in Monte Gargano'). Nuremberg 1968.
22. Emil Bock as above.
23. Walter Johannes Stein: *The Ninth Century—World History in the Light of the Holy Grail.* Temple Lodge Press 1991.
24. Reinhard Wagner: *Das achte ökumenische Konzil von 869* ('The Eighth Ecumenical Council of 869'). Die Drei 1965/1.
25. Rudolf Steiner: *Karmic Relationships*—Vol. VIII. Lecture given in Torquay.
26. News-Sheet. No. 15, 24/25 and 27/28.
27. Rudolf Steiner: *Mitteleuropa zwischen Ost und West* ('Central Europe between East and West'). GA 174a. Lecture 1 (The Spiritual background to the Outbreak of War).
28. News-sheet, No 15, p. 7.
29. ibid p. 9.
30. ibid p. 8.
31. ibid p. 9.
32. Documents. Vol. II, No. 401.
33. News-sheet No 15, p. 13.
34. ibid p. 14.
35. ibid No. 24/25, p. 23.
36. ibid p. 30.
37. ibid No 27/28, p. 27.
38. ibid p. 59.
39. ibid p. 56.
40. Documents. Vol.I, No. 40.
41. ibid No. 49.
42. ibid Vol. II, No. 241.
43. ibid No. 399 a–d.
44. ibid No. 403.

45. ibid No. 440.
46. ibid Vol. III, No 720.
47. ibid No. 717.
48. ibid No. 730.
49. Ebhard von Vietsch: 'Der Kriegsausbruch 1914' ('The Outbreak of War 1914'). In: *Geschichte in Wissenschaft und Unterricht* ('History in Science and School-Lessons'). Year 15, No. 8, August 1964.
50. News-sheet. No. 24/25 p. 23.
51. ibid No. 27/28 p. 13.
52. ibid p. 27.
53. ibid p. 31.
54. Emil Bock. As above p. 297.
55. ibid p. 300.
56. ibid p. 289.

Further reading matter

Rudolf Steiner: *Die geisteswissenschaftliche Behandlung sozialer und pädagogischer Fragen* ('The Spiritual-Scientific treatment of Social and Pedagogical questions') Lecture of 6 Feb. 1919.

Rudolf Steiner: *Fall of the Spirits of Darkness.* Lecture 14 Oct. 1917. Rudolf Steiner Press 1995.

Johannes Haller: *Nicholas I and Pseudo-Isidor.* 1936.

Barbara Tuchmann: *August 1914.* Berne 1965.

Pierre Renouvin: *Die Kriegsziele der französischen Regierung 1914–1918* ('The war-aims of the French Government 1914/18'). In: *Geschichte in Wissenschaft und Unterricht* ('History in Science and School-Lessons') Year 17, No. 3.

Ed. T.H. Meyer: *Light for the New Millenium.* Rudolf Steiner Press 1997.

Bibliography

(Books referred to in the main text.)

By Rudolf Steiner:
'Address to Young People' published as: *Youth Search in Nature*, single lecture given on 17 June 1924, Mercury Press 1984
Agriculture (referred to as 'the Agricultural Course'), Bio-dynamic Farming and Gardening Assn. 1993
'Astronomy Course', not translated in English. German edition: *Das Verhältnis der verschiedenen naturwissenschaftlichen Gebiete zur Astronomie*, Rudolf Steiner Verlag 1997
Calendar of the Soul, Anthroposophic Press 1982
'The Drama Course' published as: *Speech and Drama*, Anthroposophic Press/Rudolf Steiner Press 1986
The Etherisation of the Blood, Rudolf Steiner Press 1980
Four Mystery Dramas, Rudolf Steiner Press 1997
Karmic Relationships, Volumes I–VIII (referred to as 'the Karma lectures'), Rudolf Steiner Press
Occult Reading and Occult Hearing, Rudolf Steiner Press 1975
'The Threshold of the Spiritual World' published as: *A Road to Self Knowledge and The Threshold of the Spiritual World*, Rudolf Steiner Press 1975
Towards Social Renewal, Rethinking the Basis of Society, Rudolf Steiner Press 1999
Verses and Meditations, Rudolf Steiner Press 1979

Edited by T.H. Meyer
Light for the New Millennium, Rudolf Steiner's Association with Helmuth and Eliza von Moltke—Letters, Documents and After-Death Communications, Rudolf Steiner Press 1997

**FICTION HOUSE PRESS
PRESENTS**

AF078903

SATELLITE SCIENCE FICTION

February 1957
Vol. 1, No. 3

This reprint edition is a facsimile of the original digest magazine. Variations in printing and quality can be attributed to the original magazine which was printed on rough woodpulp paper.

New Material
© 2022 Fiction House Press

ISBN 978-1-64720-487-7

www.FictionHousePress.com
fictionhousepress@gmail.com

THE THREE-IN-ONE NOVEL

To start 1957 off with a real rocket-burst, we are running, for our complete novel in this issue, one of the most intriguing combinations of ideas and authors that, to our best knowledge, even that most fascinating of literary fields, science fiction, has yet known. It is a real triple-threat of a novel. In fact, it can be read three ways.

It began when Hal Clement, the Milton Academy geologist and author of such science fiction classics as *Needle* and *Heavy Gravity Planet*, submitted a short novel describing the actions and reactions of one of his unmatched alien entities during a brief stay on Earth. It was a great story, a terrifying story—but, at the length submitted, it was a story we could not publish in our page limits.

We wanted to publish it—but how? From a nameless source and an editorial conference emerged the idea of writing a counter short-novel, showing the actions and reactions of the Earthfolk involved toward the strange visitor. Mr. Clement's teaching activities forbade him taking on the job immediately, so we turned it over to Sam Merwin Jr.

The novel that has resulted is a contrast in styles—for two authors of more contrasting temperament, talent and experience would be hard to find. But, for once, it is a story where contrast in styles is a binding, rather than a diffusive influence. You can read it three ways, simply by perusing alternate chapters—Hal Clements alien novel skipping from first, to third and so on—or Sam Merwin's human story, taking on the *even* chapters. But PLANET FOR PLUNDER is a fine, integrated novel that stands on its own two feet.

LEO MARGULIES,
Publisher

SATELLITE
science fiction

FEBRUARY, 1957   Vol. 1, No. 3

A COMPLETE NOVEL

PLANET FOR PLUNDER
by HAL CLEMENT and SAM MERVIN JR.

The alien's mission was grim, purposeful and scientific. It was only natural that he should mistake men for machines and machines for men. Earth's terrible predicament was not of his making, but it came close to destroying Man.

.... 4 to 97

SHORT STORIES

THE LAST WORD
    by DAMON KNIGHT 98

THE NEXT TENANTS
    by ARTHUR C. CLARKE 103

FOOD FOR THE VISITOR
    by JOHN VICTOR PETERSON............... 112

THE ATTIC VOICE
    by ALGIS BUDRYS 119

FEATURE

THE SCIENCE FICTION COLLECTOR
    by SAM MOSKOWITZ 115

LEO MARGULIES
Publisher and Editor

CYLVIA KLEINMAN
Managing Editor

JOAN SHERMAN
Production

•

ALEX SCHOMBURG
Cover Art

SATELLITE SCIENCE FICTON, Vol. 1, No. 3. Published bi-monthly by Renown Publications, Inc., 501 Fifth Avenue, N. Y. 17. Subscriptions, 6 issues $2.00; 12 issues $4.00; single copies 35¢. Application for second-class entry pending post office, N. Y., N. Y. Places and characters in this magazine are wholly fictitious. © 1956 by Renown Publications, Inc. All Rights reserved. FEBRUARY 1957, Printed in U. S. A.

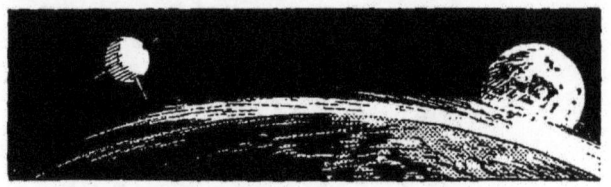

Make 1957 Your Own Man-Made Satellite Year

If you are a science-fiction reader, you already know SATELLITE. If you have not yet discovered the imaginative excitement that awaits you in reading clear, understandable stories of space-travel, of trips in time, of speculative studies of society in the years or centuries to come, you'll find it in this magazine.

You can begin your science-fiction career with SATELLITE, *not only because it is the only current SF magazine to contain a complete, self-explanatory novel, but because its stories, long and short, are selected for readability and reader-interest, rather than for small-group satisfaction. Make us your 1957 experiment!*

RENOWN PUBLICATIONS, INC.
501 Fifth Avenue, New York, 17, N. Y.

Kindly enter my subscription to **SATELLITE SCIENCE FICTION** for 6 issues @ $2.00, 12 issues @ $4.00. Remit by check or money order.

NAME ....

ADDRESS ....

CITY, ZONE, STATE ....

Please print   SSF 13

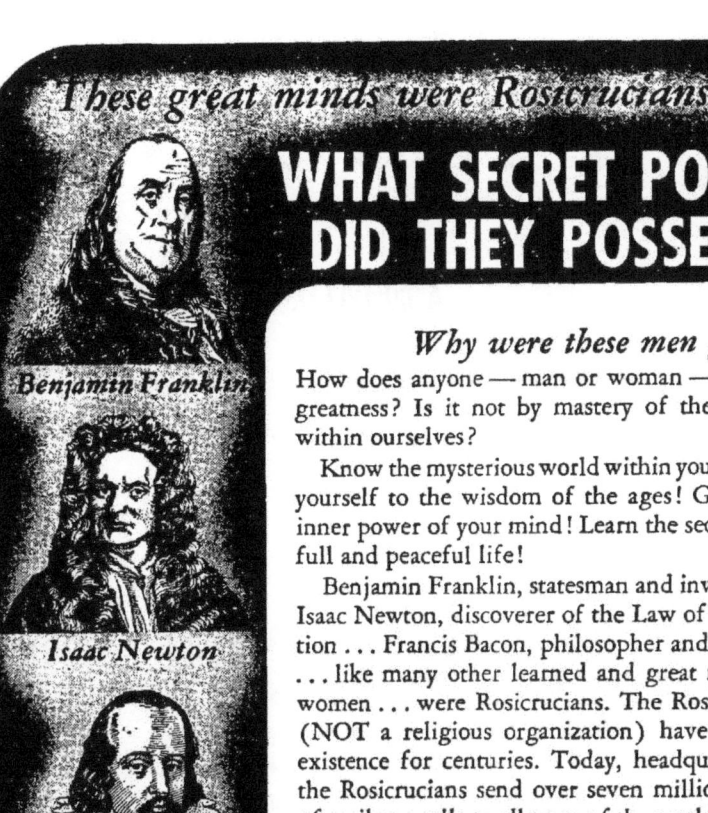

These great minds were Rosicrucians

WHAT SECRET POWER DID THEY POSSESS?

Benjamin Franklin

Isaac Newton

Francis Bacon

Why were these men great?

How does anyone — man or woman — achieve greatness? Is it not by mastery of the powers within ourselves?

Know the mysterious world within you! Attune yourself to the wisdom of the ages! Grasp the inner power of your mind! Learn the secrets of a full and peaceful life!

Benjamin Franklin, statesman and inventor... Isaac Newton, discoverer of the Law of Gravitation... Francis Bacon, philosopher and scientist ... like many other learned and great men and women ... were Rosicrucians. The Rosicrucians (NOT a religious organization) have been in existence for centuries. Today, headquarters of the Rosicrucians send over seven million pieces of mail annually to all parts of the world.

The ROSICRUCIANS
San Jose   (AMORC)   California, U.S.A.

THIS BOOK FREE!

Write for your FREE copy of "The Mastery of Life" — TODAY. No obligation. A non-profit organization. Address: Scribe X.P.S.

Scribe X.P.S.
The ROSICRUCIANS
(AMORC)
San Jose, California, U.S.A.

SEND THIS COUPON

Please send me the *free* book, *The Mastery of Life*, which explains how I may learn to use my faculties and powers of mind.

Name_____

Address_____

City_____

State_____

Planet for

A COMPLETE SCIENCE FICTION NOVEL

Out of the star gulfs he came, troubled, searching, with a warning for Earth no one dared ignore. Never would Earth see his like again—or know the reason why!

by HAL CLEMENT and
SAM MERWIN JR.

A CONSERVATION SERVICE vessel is quite fast and maneuverable as craft of that general type go. But there was little likelihood that this one would catch up with its present target. Its pilot knew that. He had known it since the first flicker of current in his detectors had warned him of the poacher's presence. But with the calm determination so characteristic of his race, he made the small course-correction which he hoped would bring him through the target area at action speed.

The correction had to be small. Had the disturbance been far from his present line of flight, he would never have detected it, for his instruments covered only a narrow cone of space ahead of him. Too many pilots in the old days, with full-sphere coverage, had been unable to resist the temptation of trying to loop back to investigate disturbances whose source-areas they had already passed.

At one-third the speed of light, such a reversal of course would

Plunder

have wasted both energy and time. No one could make a reversal in any reasonable period, and, certainly, no poacher or other lawbreaker was going to wait for the maneuver to be completed.

Even as it was, this pilot's principal hope lay in the possibility that the other vessel would be too preoccupied with its task of looting to detect and react to his approach in time. Detection was only possible if, like his own ship, the poacher carried but a single operator. Unfortunately, a freighter was quite likely to have at least two, even on a perfectly legal flight, and the Conservation pilot had known of cases where poaching machines had had crews as large as four.

Even the presence of two would render his approach almost certainly useless, since the loading and separating machinery would require only one manipulator, and the full attention of any others could be freed for lookout duty. Nevertheless, he bored on in, analyzing and planning as he traveled.

The poacher was big—as big as any he had ever viewed. It must have had a net load capacity of something like a half billion tons—enough to clean the concentrates off a fair-sized planet, particularly if it also boasted adequate stripping and refining apparatus. There was no way of making certain about this last factor, for no such equipment was drawing power as yet. And that, in a way, was peculiar, for the poacher must have been in his present position for some time.

Had the driving energies of the poacher been in use, the Conservation ship would have detected them long before, and would have experienced less difficulty in making the necessary course-change. With a scant five light-years in which to make the turn, the acceleration needed for the task was rather annoying. Not that it caused the pilot any actual physical discomfort. It was purely an emotional matter. His economy-conditioned mind was appalled by the waste of energy involved.

Four light-years lay behind him when the poacher reacted outrageously. For the barest instant the attacker dared to hope that he might still get within range. Then it became evident that the giant freighter had seen him long before, and had planned its maneuver with perfect knowledge of his limitations.

It began to accelerate almost toward him, at an angle which would bring it safely past. It would sweep past just out of extreme range if he kept on his present course—and probably well beyond trustworthy shooting distance, if he tried to intercept it. For an instant, the agent was tempted. But before a single relay had clicked in his own small craft he remembered what the poacher must already have known—that the planet, which had perhaps already been robbed, came first.

It *must* be checked for damage, even though it was uninhabited as far as anyone knew. The mere fact that the poacher had stopped there meant that it must have something worth taking. It must, therefore, be tied as soon as possible into the production network whose completeness and perfection was the only barrier between the agent's race and galaxy-wide starvation.

He held his course, therefore, and broadcast a general warning as he went. He gave the thief's specifications, its course, as of the last possible observation, plus the fact that it seemed to be traveling empty. The absence of cargo was an encouraging sign. Perhaps no damage had been done to the world ahead. Unfortunately, it might also mean that the raider had a higher power-to-mass ratio than any freighter the agent had ever seen or heard of. But that he seriously doubted. He assumed that the ship was without cargo, and worded his warning accordingly.

His temper was not improved by an incident which occurred just before the giant vessel passed beyond detection range. A beam, quite evidently transmitted from the fleeing mass of metal, struck his antenna, and the phrase—"Now, don't you just hope they'll get us!"—came clearly along the instrument.

Again, relays almost closed on the Conservation flier, but the agent contented himself with repeating his warning broadcast and adding to it the data which had inevitably come along with the poacher's taunt—data concerning the personal voice of the speaker. Then he turned his attention to the problem of the planet ahead.

He would need more energy, of course. The interstellar speed of his craft had to be reduced to the general velocity of the stars in this part of the galaxy, for he could not make the survey that would be needed, merely by viewing the planet as he flashed by. He could, of course, get a pretty good idea of the metals that were present through such flash-technique, but he needed information as to their distribution. If he were lucky—if the poacher had actually failed to load up—there would almost certainly be concentrates worth recording and reporting to Conservation.

The sun involved was obvious enough, since it was the only one within several light years. The agent thought fleetingly of the loneliness, even terror, which would descend upon the average ground-gripper in close proximity to the nearly empty space at the galaxy's rim and timed and directed his deceleration to bring him to rest some twenty-four diameters from the sun's photosphere.

The poacher had begun to travel long before he drew close enough to detect individual planets, and he was faced with the problem of discovering just which planet or planetoid had been visited. There were certainly enough to choose among and he was reasonably sure he had detected them all as he approached.

The possibility that he had been moving directly toward one for the whole time, and had, as a result, failed to observe any apparent motion for it, was too remote to cause him concern, particularly since it turned out that he had been well away from the general orbital plane of the system. He had the planets, then. But which ones were important? Since he would have to check them all anyway, he didn't worry too much about selection. After using up the energy and time needed to stop in this forlorn speck of a planetary system, it would be senseless to leave anything unexamined. Why, he reasoned, should anyone else have to come back later to do what he had left undone? Still, he thought, it would be pleasant to determine quickly what the poacher had accomplished, if anything.

The innermost planet was definitely not the plundered victim. It had plenty of free iron, of course, and the agent noted with satisfaction that the metal was not concentrated at its core. If it ever became necessary to seek iron so far out in the galaxy, stripping it from so small a world would be relatively easy.

However, the important metals seemed to be dissolved and distributed with annoying uniformity through the tiny globe—a fact which was hardly surprising. The planet was too small, and its temperature was too high to permit either water or ammonia to exist in liquid form. The ordinary geological processes which produced ore deposits simply could not function here.

The second world was more hopeful—in fact, it seemed ideal on first survey. There was water, though not in abundance. Nevertheless, in the billions of years since the planet had formed a certain amount of hydrothermal activity had gone on in its crust, and a number of very good copper, silver, and lead concentrations appeared to exist. The agent decided to land and map these, after he had completed his preliminary survey of the system. If this were the world the poachers had been sweeping, they had evidently failed to get much. Venus might be the plundered planet.

It proved not to be, however. Earth's water is not confined to its lithosphere—it covers three-quarters of the planetary surface. It washes mountains into the seas, freezes at the poles and, at high elevations, even at the equator. It finds its way down into the rocks and joins other water molecules which have been there since the crust solidified. It picks up ions, carries them a little way, and trades them for others.

In short, Earth contains enough water to produce geological phenomena. The agent saw this almost in his first glance. He wasted a brief look at the encircling dry satellite, then he turned all of his attention on the primary planet itself. He even began to ease his ship outward from the orbit it had taken up, twenty million miles from Sol.

This, he decided, must be the world of the poacher's selection. Even without analysis, anyone with the rudiments of a geological education would know that there must be metal concentrations here —and a civilization that uses half a trillion tons of copper a year can be expected to have at least a few trained geologists.

The agent pointed the nose of his little cruiser at the tiny disc, shining brightly eighty million miles away. He drove straight toward it, combing its surface as he went with the highest-resolution equipment he could bring to bear. All over the surface, and for a mile below, those radiations probed and returned with their information. The agent swore luridly as the indicators told their tragic story.

There *had* been concentrations, all right. There were still a few. But someone had been scraping busily at the best of them, and had left little that was economically worth recovering. It was the old story. If good deposits and poor ones were worked at the same time, the profit was of course smaller. But at least the deposits lasted longer.

An eternity had passed since any legal operator of the agent's race had worked the other way, stripping the cream for a quick profit and letting the others go. Such a practice would have crippled the industry of the agent's home planet millions of years before, had it not been checked sternly by the formation of the Conservation Board.

Crippled industry, to a race at the stage of development his had attained, was the equivalent of a death sentence. Not one in a thousand of his people could hope to escape death by starvation, if the tremendously complex system of commerce were to break down.

The agent knew that—like most of his profession, he had seen border worlds where momentary imperfections in the system had taken their toll.

His fury at the sight of this planet mingled with—and was fed by—the memory of the horrors he had seen. Apparently, he had been wrong. The poachers had gotten away with their load —in fact, scores of them must have been at work.

No one ship, not even the monster he had seen so recently, could have done such a job without assistance on a planet of this size. The Conservation Department had suspected, before now, that it faced a certain degree of organization among the poachers. Here was infuriating evidence that the suspicion was all too well-founded.

Thought followed reaction through the agent's reception apparatus and through his mind, before his ship was within a million miles of the planet.

At that range no precise mapping was possible. In a sense, surface-mapping was no longer necessary, since the surviving deposits were hardly worth the gathering—but the tectonic charts would have to be obtained as usual.

A world like this was in constant change. A million, or ten, or a hundred million years from now the natural processes within its crust would have brought new concentrations into being. These forces must be charted, so that proper predictions could be obtained. Only through such research and predictions could Conservation beat the poachers to the next crop of metal, when it appeared.

The agent began to decelerate again, now matching his velocity with that of the planet itself. At the same time, he began a more detailed analysis of the surface, refining it constantly as the distance diminished. The water he already knew about. He had supposed the gaseous envelope to consist of methane and water vapor, with perhaps some ammonia, formed at the same time as the rest of the planet. But his instruments told a different story.

Earth had lost its primary atmosphere. The tragedy had occurred before the first member of the agent's race had ventured away from his own planetary system. The agent found the free oxygen, and swore again. He knew what that meant—*photosynthesis*. The planet was infected by those carbon compounds that behaved almost like life, except for their ferociously rapid rate of reaction.

They were not very dangerous, of course, but due care had to be exercised, and constant vigilance maintained. A good many planets in the liquid-ammonia-liquid-water temperature range had them, and techniques had long since been worked out for conducting analysis, and even for mining in their presence, destructive as they often were to machinery.

The Conservation vessel, naturally, was constructed of alloys reasonably proof against any attack by free oxygen or the usual run of the carbon compounds. In fact, if this world had any unique developments of the latter the agent could always lift his ship out of the atmosphere. Such a retreat seemed to put a stop to the growth of photosynthetic life.

It never occurred to the agent that concealment might be in order. In the first place, he was on a perfectly legal mission. In the second, equally of course, he didn't think that there might be anyone on the planet to observe his arrival.

Oxygen being what it is, he had automatically classified the world as uninhabited and uninhabitable. As a result, the events of the half-second following his machine's penetration of Earth's ozone layer demanded a rather drastic revision of his outlook.

The radar beams, for an instant, made him suppose that another ship was on this world, and was trying to communicate with him. He had almost begun to answer before he realized that the radiation was not modulated, and could hardly be speech—or, more accurately, that its modulation was too simple and regular to represent words. Even though such radiation did not mean intelligence, however, it obviously did imply the presence of life.

Somehow, an organism must have evolved in an oxygen atmosphere with the ability to reduce metal oxides or sulfides, and keep them reduced to free metal. At the moment, it seemed to be a low order of life. But if it continued to develop as the agent's own species had done, this corner of the galaxy might become rather an interesting place in time. A man might have drawn a somewhat similar conclusion from hearing the chirp of a cricket under analogous circumstances.

At first, the agent supposed the radiation to have meaning similar to that of the cricket's chirp, too —it came and it went, regularly and monotonously, from a seemingly fixed source, and had an apparent willingness to go on until the sun cooled. But, a few milliseconds after the first pulses struck his receptors, others began to come in. They shared the simplicity of pattern shown by the first, but there were more of them.

As the ship moved, and its distance from some of the sources changed, it became evident that the waves were being directed in beams, rather than broadcast in all directions—and that the beams were following the ship. Intelligent or not, *something* was at least aware of his presence.

A score of hypotheses ran through the agent's mind during the next few milliseconds, for thought can move rapidly, when the neurons involved are of metal, and the impulses they carry are electronic currents, rather than potential differences between the surfaces of a colloid membrane. But none of these theories managed to satisfy him.

Even he could not continue to theorize at the moment, either— for the hull of his vessel was glowing bright red, and the surface of the planet was coming up rather rapidly to meet him. He had to land within the next few seconds, assuming that he did not want to do his theorizing hanging motionless in the atmosphere.

The outer surface of his hull was a trifle hard to manage at its present temperature. But none of the myriads of relays further in had been affected and the sliver of metal obeyed his thoughts as it always had, slowing to a dead halt a few yards above the surface and then settling down for a landing while the agent analyzed the material directly underneath.

It was pure luck that there was no vegetation below him—luck, at least, for any local fire-fighters. The hot walls did respond to control, albeit a trifle sluggishly. Particles of sand and clay, coming in contact with the hull began to dance, like bits of sawdust on a vibrating plate. And like sawdust, the dance carried them into a particular pattern.

The pattern took the form of a hollow under the hull, while the excess soil heaped up around it on all sides. The ship eased gently downward into the crater thus formed, which deepened as it continued to sink. The settling of the vessel, and the deepening of the hole, continued for perhaps twenty feet, before the hull touched solid rock.

When it did, more relays moved, and the rock itself flowed away in fine dust. This continued for only another foot, and then the ship was resting in a perfectly fitting cradle of stone, and the displaced soil was drifting back around it, covering its still red-hot circumference. The sand smoothed itself into a low mound which almost, but not quite, covered the vessel.

Had the agent cared about concealment, of course, he could have dug a little deeper—but all he wanted was good contact with bedrock. There was much mapping to do, and the matter of local life would have to wait until it was done.

II

IT WAS ONE of the new, triangular, floating radar installations, some two hundred miles northeast of the Virginia Capes, that first picked up the track of the interstellar visitor. Since the vessel was still well up in the Photosphere, far too high for even the latest model planes, the report was worded, . . . *Unidentified Flying Object, altitude (tent.) 50 mi. plus, speed (tent.) 3,600 mph, direction northwest by west* . . .

The ever-watchful, and super-sensitive network, set up to guard the lives and property of a continent, responded with an instant alert. In the central communications hall of a huge building near Washington, D. C., worried experts and officials gathered around plotting boards or stood in tight-lipped silence before a gigantic map on which reports were automatically registered in moving beams of light. The International situation was hardly tense enough to make probable an immediate enemy action. But in a Cold War period there could be no let-up of suspicion or instant readiness to act.

"The damned thing, whatever it is, is headed straight for Chicago," growled a grey-haired brigadier general, whose face was seamed and leathery from hundreds of air-combat hours.

"She's coming down, too," replied a civilian expert, frowning at the latest reports, which were coming in with increased rapidity as the strange aerial object swept over thickly populated sections of the country. "Altitude only thirty miles over Akron. And she's losing speed by the minute."

"That's what worries me," replied the brigadier unhappily. "If it were a meteor it would be picking up speed. It would be blazing like a comet, even in broad daylight."

"Nobody has yet developed a long-range missile control that will brake an enemy aircraft over the target," said a third member of the high-echelon group, one who wore the light grey-blue of a naval officer in summer uniform. He spoke quietly, almost shyly, but his chest, beyond a highly-colored array of battle and medal-ribbons, carried the heavy silver wings of a command pilot.

"I don't like it," said the brigadier, thrusting his hands deep in his trouser pockets. "Just because we don't have this sort of long-range missile control, doesn't mean that *they* haven't come up with it. All those scientists they've been turning out—and the hotshots they grabbed when they moved into Germany, in nineteen forty-five—" He let it hang there.

"Lord knows, meteors *have* been known to act freakishly," said the naval air officer.

At this point, Great Lakes Station came in with a report that put the UFO, still slowing, still descending, at a point well west of Chicago. There was a general sigh of relief.

But the brigadier remained unhappy. "We'll have to alert every interceptor group in the Northwest," he said quietly. "At the rate that mystery crate is coming down, we'll be able to track up after it any minute now—shooting."

It was the civilian who voiced the thought that had been in all their minds—the thought which none of the others had dared to put into words. He said, "That's going to do us a hell of a lot of good, if she turns out to be a flying saucer. She'll simply take off and zoom out of range."

Nobody answered him, though long looks were exchanged. Then they all went back to checking reports, to planning the interception that seemed to grow more possible with each passing minute. The path of the object seemed to be turning more directly west as its speed continued to lessen, and its altitude to abate. Interceptor command groups within range of its path were ordered to stand by for scramble. Unfortunately, as the object came within

Nike range, it was in a part of the continent where no rocket interceptors had been installed.

Then came a phone call from more than 2,000 miles away—from the lips of the general commanding the nation's Intercontinental Bombing Command. In accordance with their routine of constant test-missions, a squadron of B-52's, much too high for civilian observation, had been carrying out an overnight mock-bombing flight from its home field, in Texas, to a uranium mining complex far up in Northwestern Canada, near Great Slave Lake. Currently, they were making their return journey back to Texas.

Said the commanding general, his voice curiously crisp despite its nasal Midwestern drawl, "Three of my observers just spotted your UFO, flying a course a few points north of due west. It was two miles above them, moving at more than fifteen hundred. It was round and red-hot."

"You mean round—like a saucer?" the brigadier asked, his voice breaking.

"No—it was round like a cannonball. And hotter than an H-bomb!" was the response.

When he had hung up, Minneapolis came in. Object safely past, still descending, still losing speed . . . Bismarck, North Dakota, had the object heading due west. Then came a ground observer report from Miles City, Montana, and another from Billings. In both cases, it had been seen as a round, red-hot object, streaking westward across the sky.

Then, nothing . . .

It was a rough day for the Radar Network.

IT WAS ALSO, as events were to confirm, a rough day for Field Expedition Seven, Summer of 1957, Montana University of Mines, Departments of Geology and Climatology.

Measured by its human components the expedition was a modest one and consisted of Assistant Professor Harold Parsons, his wife, Candace, and a Climatology Fellow, and Field Worker Donald MacLaurie, known to the regional sportswriters as *Truck*.

Their equipment consisted of one jeep with two-wheel trailer, two tents that had just been stowed away for daytime travel, canned food supplies, and an assortment of tools and instruments, including a Geiger counter bootlegged by Truck MacLaurie and currently the subject of argument between Truck and Professor Parsons.

"Listen, Truck," Parsons said, with all the patience he could muster. "This is a university field expedition, not a uranium hunt. If you want the credit you'll need to play football this fall, you'll keep that click-box out of sight and out of mind. We're here in the hills to study variations in surface clues to copper-ore formations—that is, I am here for that purpose. With your help, of course—if help is just the proper word for it. Candace is here to study cloud formations in the hills, for long-range precipitation effect on mining operations at Butte and Anaconda. I'm hoping you'll learn enough about geology to enable me to give you that credit, come September—without putting a permanent mortgage on my professional integrity."

"Golly, Doc—I only intended to try her out during my spare time," protested Truck.

"What I'm trying to say, Truck, is this. There isn't going to *be* any spare time on this trip." Parsons paused, and added with a trace of acid, "You're not back sleeping in classroom now. *You're in the field!*

Parsons didn't have to look at his wife to know how she was reacting to his lecture. Not that Candace would show disapproval in the presence of an outsider. But he was all too familiar with the slight blankness of usually alert and sympathetic brown eyes, the invisible aura of coolness that surrounded her. There were moments when he wished she weren't quite so sympathetic and outgoing in her relations with people. It only made his own diffidence more pronounced.

Nor was he helped by the fact that, though he stood a wiry six feet one in his socks, he had to look upward to meet Truck MacLaurie's large and blandly childish blue eyes. He also felt hampered by the fact that, while he himself was close to thirty, and Truck a mere twenty-two, the big ox looked about five years his senior.

He was about to cut it short and say, "All right, let's get started,"—when the UFO passed, whizzing, over their heads.

It could not have been more than a mile above them, and it was round as a gigantic egg from some monster bird, red hot as a cooking stone in some giant's barbecue-pit. It was traveling like a bat out of hell, due west, and it was falling fast. Even at that distance, it left in its wake a lingering sense of tremendous heat.

"Golly!" said Truck, following the object's progress with open disbelief. "It's gonna crash that crummy hill, head-on!"

The expedition of three had made camp, the night before, close to the center of an arid valley in the eastern foothills of the Rockies, roughly halfway between the mining communities of Brown and Hamilton. And camped there they still were, in a district where even the decaying remnants of ghost mining communities were scarce. It was rough, wild country—about as rough and wild as Rocky Mountain foothill country can ever get.

The western end of the valley was blocked by a range of minor hills whose topmost peak rose no more than five thousand feet from the valley floor. Unerringly, the speeding object appeared headed for this peak. Looking on with a mixture of amazement and disbelief which precluded horror, Parsons tried to remind himself that perspective played strange tricks, and that the object, whatever its nature, was undoubtedly on a course that must carry it hundreds, perhaps thousands, of miles before it crashed into the rugged terrain.

Then, unaccountably, the object swerved to the south, avoiding collision as neatly as a plane skillfully piloted by a crack ace. It disappeared *around* the peak, not *over* it, and vanished from sight behind the ragged mountain wall.

Then, there was nothing . . . no crash, no explosion. Nothing at all!

"Hal honey," said Candace Parsons, "will you, for the love of Osiris or whatever gods you worship, light me a cigarette?"

Not another word was spoken for almost two minutes. The three of them stood there, spellbound, staring at the wall of hills, waiting for something, for anything. But there was nothing.

Again, it was Candace Parsons who broke the silence. She was a trim, long-legged girl, with soft brown hair with a texture so fine that it defied shop and home permanents alike. She was remarkable, too, in that her figure and appearance remained pleasantly female, despite her all-over ranginess and the disfigurement of camping clothes.

She said, "Since neither of you geniuses has any idea of what it is, I think we ought to report it, don't you?"

Parsons nodded. He stepped on his own cigarette and ground it out in the sandy soil. "Perhaps if that damned transmitter of ours can clear those hills we came through yesterday . . ." He let the sentence trail off, and with Truck MacLaurie went back to the jeep.

The two of them broke the radio out of the trailer and set it up in the open. After fifteen minutes, it became clear that they were not going to get through. Parsons disconnected the transmitter and nodded to Truck to cease winding the battery. He looked at the football player almost pleadingly.

"No, Doc," said Truck. "If you think I'm gonna wheel this buggy back over the hills while you and Candace have all the fun . . . Well, the answer is no. Let the credits fall where they will."

"Why, Truck!" exclaimed Candace, who had taken a Bachelor of Arts degree in English at a

Midwestern university one year before her interests had veered to Parsons and Climatology. "That's almost poetical."

She saw the way both men were looking at her and shook her head. "I'm not going back either," she added firmly. "I've always wanted to look at a UFO, and if you think I'm passing up this chance—"

Parsons squinted at the hills ahead. He said, "Okay, Bounty mutineers. Let's put this show on the road. We can run up the transmitter when we hit the next range of hills, and maybe get a message through to Hamilton or Stevensville."

A moment later, as they took their places in the jeep, he asked: "Candace, you wouldn't kid me, would you?"

"Who's kidding?" she countered. "That thing didnt look like a flying saucer, but it didn't look man-made either. And who ever heard of a meteor with sense enough to detour *around* a mountaintop?"

"Maybe it's a good thing I brought the click-box along after all," said Truck, who was massively filling the rear seat of the

jeep. "You can't tell what that thing may be radiating when we find it."

"*If* we find it!" Parsons said quietly, steering the rugged little vehicle neatly around a treacherous rock outcropping that lay concealed by a mask of brush.

The object, whatever it was, had flashed over the valley in less than a minute. Covering the same distance by ground had taken the expedition almost all day.

Throughout the morning, and early afternoon, the hot summer sun beat unmercifully down upon them out of a pale blue sky, reddening already painful sunburns and causing Truck MacLaurie to break out with a rash of prickly heat that had him scratching himself almost continually.

They made only a brief stop for lunch—under a low hill that offered a very poor kind of shade. They would not have stopped at all, but Candace insisted they needed to stretch their legs even if they had lost their appetites. She served them slices of processed ham on crackers, and coffee so hot that Truck wanted to know whether she had actually used the heater or had merely left it out in the sun for two or three minutes.

When they resumed their journey, they found the sun shaded by freshly-assembled clouds, which caused the big football player to mutter, "Thank God! We've got the weather on our side, anyway."

"I don't like it," said Candace, staring thoughtfully at the western sky ahead. The jeep was bumping its way up toward the mountains, and something in her tone caused Hal Parsons to slow down and look at her sharply.

"What is it, baby?" he asked.

"According to meteorological tables, there has been an average of only one inch of rainfall in these hills during August," she said. "Furthermore, there has never been a *single* rainfall of more than a quarter of an inch. Those clouds piling up ahead spell out heavy rain to me."

"Speaks the climatologist," said Parsons, concentrating on a barren, rocky patch of hillside ahead.

"Speaks a girl who'd like to know what's going on," Candace replied.

In the rear seat, Truck MacLaurie scratched, sweated and said nothing.

Slowly, circuitously, Parsons and Truck got the little jeep and its trailer up the roadless hillside, and set its course toward a notch in the hills ahead. It was after five o'clock when they reached the summit of the pass and could look down into the valley on the other side of the range. To all intents and purposes, it was an exact replica of the valley from which they had so painfully emerged.

"I don't see anything," said Truck, letting his eyes roam the panorama of scrub and sand and eroded rock that stretched out before them.

"I don't either," said Parsons, battling a feeling of disappointment which he knew to be absurd. There was no reason to believe that a flying object capable of detouring the mountain peak on their right would have selected such a barren piece of earth as its resting place. It could just as easily have steered past other peaks, over other valleys, until it reached a wide, fertile valley.

"Look at those clouds!" said Candace Parsons, staring not at the valley or the range of hills opposite, but at the sky above. "If we don't get ourselves set up quick, we'll be in for a wetting."

"Hello!" Parsons exclaimed, following his wife's gaze. "It does look grim—and it seems to be right overhead. Come on, Truck."

"I'm glad we're up here, instead of down there," said the king-sized young gladiator, moving to help his professor with the camping gear. "Only time I ever saw clouds like that was in Colorado, when I played left guard for a junior college. We wound up with a flash flood that wiped out half the campus. I came up to Montana Mines the next year."

"Better give the radio another run, before you break out the tents, Hal honey," Candace Parsons suggested. "I'll take charge of them until you get a message through."

Even as she spoke, a large drop fell on her forehead, and the slow patter of beginning rain rustled around them. She got busy with the tents, moving swiftly, efficiently, and with surprising strength for a slim-looking woman. But her mind was on the weather that was encompassing them.

It simply couldn't be happening—but it was. Great grey-and-white, swirling masses of cloud had boiled over to fill the heavens above them, and the fall of rain was increasing steadily. Her experience told her that these were not cloudburst formations, from which only a flash flood could be expected. Despite Truck's concern—they looked more like the prelude to long, steady rain. Yet they were low, closing down relentlessly, making the ceiling almost invisible as they blanketed the taller peaks that rimmed the valley.

It was her husband who broke through her abstraction, saying, "Baby, come here. I got through to Stevensville. According to them, every watcher and radar post has been alerted all day. And we're the furthest west observers to have seen that thing. They've only been waiting for another report to start an air-search."

"If it landed anywhere around here," said Candace Parsons, eyeing a sodden cigarette in disgust, "they won't be able to get a plane over these peaks for quite a while —not even a helicopter."

III

THE RADAR BEAMS had stopped —or had, at least, ceased to reach the Conservation Agent—before he had gone underground. The point where he had landed was not in line-of-sight range of any of their stations. Needless to say, however, their operators had not forgotten him.

The agent was not considering possible radar operators. In fact, he would not have considered them even if radar had been, to him, something produced by a machine. He was far too busy listening.

If a human being puts his ear close against a wall, or a doorjamb in a fairly large building he will pick up a remarkable variety of sounds. He will hear doors closing, windows rattling, and assorted creaks and thuds whose origin is frequently difficult or impossible to determine. The one thing he will not hear is silence.

The crust of a planet is much the same on a vastly greater scale. It is always full of vibrations, ranging from gigantic temblors— as square miles of solid rock slip against similar areas on the two sides of a fault plane—to ghostly echoes of sound and the faintest of thermal oscillations as the sun's heat shifts from one side of a mountain to the other, and the rocks expand and contract to adjust to the new temperatures.

These waves travel, radiating from their point of origin, being refracted and reflected as they enter regions of differing density or elasticity, losing energy as they go by heating infinitesimally the rock through which they pass. They may die out entirely in random motion—heat—while still inside the body of the planet. Or a good, healthy wave-train may get all the way to the other side.

If it does so on Earth, it takes about twenty minutes. Then a fair proportion of it bounces from the low-density zone that is the bottom of the atmosphere, or the top of the lithosphere, whichever you prefer, and starts back again.

And every variation in density, or crystal structure, or elasticity, or chemical composition, has some effect on the way such waves travel. They may speed up or slow down. Transverse waves, or the transverse components of complex waves, may damp out—have you ever tried to skip rope with a stream of water?—and the compressional waves alone go through. Transverse waves, polarized in one direction, may be refracted through an interface, where the same sort of wave striking the same interface—at the same angle but polarized differently—may be reflected from it. The important thing is that constantly varying conditions affect the waves. And that means that the waves carry information.

It is confused, of course. Temblors come from all directions, from all distances, due to many different causes, and through all sorts of rock. Interpreting them is not just a matter of sitting down to listen. One might as well tune in a dozen different radios to as many different musical programs, while sitting in the middle of a battlefield with a thunderstorm going on, and try to decide how many flutes were being used in one of the orchestras. The information is there, but selectivity and analysis are needed.

The agent was equipped for such selectivity, such analysis. His sensitive gear could detect any motion of the rock, down to thermal oscillation of the ions, at frequencies ranging from the highest a silicate group could maintain, to the lowest harmonic of a planet the size of Jupiter.

If his instruments proved inadequate, he could listen himself. But since just listening would involve the projection of a portion of his own body through the hull and bringing it into contact with the rock, the act would put a crippling strain on his stone-like flesh, and would consume several millennia of time. He did not plan to take this alternative. Machines were built to be used. Why not use them?

His own senses reacted at electronic speed—were, in fact, electronic in nature, as were his thought patterns. The process of receiving a group of impulses, and of solving the multiple-parameter equations necessary to deduce all the facts as to their origin and transmission, called for just such a fast-acting computer as his mind, though even he took some time about it.

This, primarily, was because he was careful. A temblor originating near by would naturally have fewer unknowns worked into its waveform by the time it reached him. Therefore, it represented a simpler problem. Also, when solved, that problem provided quantities which could be fitted directly into the equations for more distant wave-sources, since their wave-trains must have come through the same rocks as they approached him.

His picture of the lithosphere around him grew gradually, therefore, and by concentric shells. He saw the layers of different sorts of rock and, far more important, the stresses playing on each layer—stresses that sometimes damped out to zero in the endless, tiny twitchings of the planet's crust and that sometimes built up until the strength of the rocks, and the vastly greater weight of overlying materials could no longer resist them, and something gave.

He sensed the change, as trapped energy built up the temperature in a confined volume, until the rock could no longer be called solid, even though the pressure kept it from being anything that could be called liquid. He saw the magma pockets formed in this way migrate, up, down and across in the crust, like monstrous jellyfish in an incredibly viscous sea.

He saw certain points on the planet where they had reached so nearly to the surface that the weight above could no longer restrain the pressure of their dissolved gases. An explosive volcanic eruption is quite a sight, even from underneath.

His senses, through the vessel's instruments, probed down toward the core of the world, where magma pockets were more frequent. In such pockets, held in solutions which might some day carry them to the upper crust, they would be accessible—the copper and silver and molybdenum, and other metals his people needed. They would lay diffused through the material of the planet.

Those were the things that interested him. He needed to know the forces at work down there—not in general, as a climatologist knows why Arizona is dry, but in sufficient detail to be able to predict when and where these metals would reach the upper crust and form ore bodies. The fastest electronic computers man has yet built would be a long time working out such problems, given the data. The agent was certainly no faster, and was less infallible.

He knew this to be so, and, therefore, spent much of his time checking and rechecking each step of the work. The task took all his attention, and, for the time being, he was totally indifferent to impulses originating near the surface —much less to a number of feeble ones which originated *above* the surface.

There was something a good deal more interesting than human reactions to claim the attention of the Conservation operative. He had, of course, confirmed long since his original impression that the ore beds of the planet had been looted. His principal job now was to decide how long the normal diastrophic and other geological processes would require to replace them.

On a purely general basis, replacement should take tens of millions of years for a planet of Earth's size and constitution. Magma pockets would have to work their way up from the metal-rich depths to the outer crust. Then they would have to come into contact with materials which would dissolve or precipitate, as might be the case, the particular metals he sought.

The geological processes which depended so heavily on water or ammonia, in the liquid state, and concentrated the metallic compounds into ore deposits, could occur only near the surface. Of course, a magma pocket, commencing five hundred miles down, may not go upward. It may travel in any direction whatever, or not at all.

The density, the chemical composition, the melting point of the surrounding material, its ability to retain, in solution, the radioactives which may have been responsible for the pocket in the first place were all vital factors. Equally vital was the question of whether its crystalline makeup is such as to absorb or release energy as increasing temperature reorganizes it—the proximity of one or more of the vast iron-pockets, whose coreward settling contributes its share of energy. All of these things influence the path as well as the very existence of the pocket.

It would be relatively easy to predict, on a purely statistical basis, the number of ore-bodies to be formed in a given ten-million-year period. But the agent needed much more than that. When a freighter is dispatched to pick up metal at one specific point and deliver it to another, the schedule is apt to suffer, if the ship has to wait a million years for its load. Interrupted schedules are not merely nuisances. In a civilization spread throughout the core of the galaxy, none of whose member worlds are self-sufficient, they can be catastrophic.

So the agent measured carefully, and, as he did so, something a trifle queer began to appear. Impulses that did not quite fit into the orderly pattern he had deduced kept arriving—impulses of a nature he found, at first, hard to believe.

Then he remembered that the poachers had been here for quite a while before his own arrival, and an explanation lay before him. The impulses were of the sort that his own hull must have broadcast, while he was digging his present refuge. There could be only one thing which the poachers would logically have left behind them. They could have left evidence of their digging.

They had shown, he decided, a rather unusual amount of foresight for their kind, coupled with a ruthlessness which made the agent wonder whether they had even felt the radar beams that had greeted his own arrival. What the poachers had done was not a thing to do to an inhabited planet.

The out-of-place impulses were from mole robots, slowly burrowing their way into the world's heart. Each one, as the agent patiently computed its position, course, and speed, was headed for a point where the release of a relatively minute amount of energy would swing delicately balanced forces in a particular direction. The direction was obvious enough. The poachers expected to be back for another load, and were stimulating Earth's diastrophic forces to provide it.

This was a technique often used by legitimate metal-producers, but only on worlds that were uninhabited. Orogeny, even when stimulated in this fashion, may take half a million years to raise a section of landscape a few thousand feet. That still would not provide time to escape for beings who, without mechanical assistance, would take something like the same length of time to travel a few hundred.

From the agent's point of view, the presence of such depth-charges meant that Earth was going to become, in a fairly short time, a writhing, buckling, seething surface of broken rock, molten lava and folding, crumpling, tilting rafts of silicate material on a fearfully disturbed sea of stress-fluid.

Such heartless behavior might prove unavoidable—since he wouldn't be there at the time. But—what had produced those radar beams?

It revolted him that any planet with life should be treated in such a manner. Whether or not the life was currently intelligent was beside the point. Few generations were needed to transform a life-species, from something as unresponsive as the planet that had spawned them, into a species capable of understanding the internal mechanism of a star in detail, for any distinctions of that nature to carry weight. If those beams had originated from living bodies, something would have to be done about the moles.

The agent simply had not the equipment to do a thing. He could fight his little ship. He could investigate and analyze. He could communicate all the way across the galaxy, if something like the ionized layers of a planet's atmosphere did not interfere.

But he had no mole robots on his vessel, no weapons that would penetrate rock, or even atmosphere, for any great distance. He could not himself stand the temperatures at depths to which some of the poacher's moles had already penetrated. Consequently, he could not follow them in his own ship, even if it were able to dig as rapidly as the robots. It was indeed a problem!

Sending for help was possible, but almost certainly useless. His patrol area was so far out near the galactic rim that any message would take several millennia to reach a point where it would do any good—and the ships which answered it would be at least three times as long in covering the distance as the radiation that summoned them.

By then most, if not all, of the robots would have reached their designated target points. They would have shut off the fields which held their shape against the pressure of the surrounding rock. Once that protection was gone, no material substance in the universe could keep the half-ton of fissionable isotopes forming their cargoes at subcritical separation. All that energy would come out, and the little that wasn't heat to start with soon would be.

Of course, even such amounts of energy are small in comparison with the usual supplies of a planet's crust. But once released in carefully calculated spots and at even more carefully calculated times they would do exactly what the poachers wanted. The Conservation agent, checking the placement of the moles, could find no fault with the computations of the poachers' geophysicist. He was in his own way an operator of genius!

He could, of course, arrange for official freighters to be on hand when the action bore fruit, and would certainly do so, as a last resort. But he must first attack the question of whether or not life was being endangered. For the first time since the beginning of his analysis, the agent directed his attention to the surface layers of the world.

Then he almost stopped again, as a new theory struck him. This planet had free oxygen in its atmosphere. Would its life, if any, be near the surface? But his hesitation was only momentary. He recalled the radar beams which were his only reason for suspecting life. They could not possibly have passed through any significant amount of rock. While his senses swept the surrounding crust, in ever-widening circles, he pondered the question of just how a living creature could endure such an environment. *Think hard now, concentrate!*

There was one obvious possibility. It might be riding a machine designed to protect it, as he was himself—which would imply that life was not native to this world. If that were the case, locating the creature or creatures should be easy. However, in such circumstances it would have to be assumed that the population was very small, since furnishing machines for all of a large population was a manifest impossibility. It would be unwise too—even if such a thing *were* possible.

A more fantastic idea was that, while the life of this world might have a carbon composition like his own, its metallic parts were of more inert substances—perhaps of the platinum-group metals. The agent knew no reason why these should not serve as well as calcium, in a nervous system. He might have thought of aluminum, had he been familiar with its behavior in an oxygen-water environment.

Then, there was the notion that a ship of his own race might be down and crippled—the most fantastic of all. No such ship would be this far out in the galaxy, and it was hard to imagine a mishap which would leave the operator alive and safe from the environment, while crippling his communication facilities to the point where nothing but crude whistles came through.

Furthermore, there had been too many points of origin for the beams that had touched him. It might prove a difficult nut to crack.

In fact, it was simply impossible to decide whether one of these hypotheses, or something which had not yet occurred to him would prove closest to the truth. For the time being, there was nothing to do but search. Naturally, it did not take long for the more or less rhythmic impulses originating only a few miles away to catch his attention.

They were seismic, of course, since he was doing all his listening through the rock—but it quickly became evident that they were originating at the very boundary between lithosphere and atmosphere. Almost as quickly, he realized that the sources were moving. This latter fact complicated the analysis rather seriously. It took the agent some time to conclude that sets of more or less solid objects, apparently always in pairs, were striking the lithosphere from outside. Sometimes there were relatively long periods of regular, repeated thuds, as one or more of the pairs did its hammering and such periods were always accompanied by motion of the point at which the blows were occurring.

At other times, the hammering was irregular, both in frequency and energy, and usually, though not always, these sequences radiated from a relatively fixed broadcasting point. There seemed to be six basic units producing the impulses. Well, he was making progress, at any rate. Systematic thought could be a joy in itself!

Quite evidently, if this disturbance were caused by local life, that life must be civilized to the point where it could design and build machines. Furthermore, six machines, machines so close together, really did call for thought. It suggested something about the population density of the planet.

On the worlds the agent knew, scarcely one individual in a thousand manned a machine capable of moving him about. To equip the rest similarly would not only be the height of folly. It would be impossible, because enough material could never be obtained, still more because very few of them were temperamentally suited to physical activity. Even if this race had equipped, say, one in a hundred of its members, the finding of such a number congregated in one spot implied either a tremendous population density or—*could it be that they were looking for him?*

He had never stopped to think what a two-dimensional search would be like. But these machines, he was beginning to think, must be confined to surface-travel—perhaps sub-surface as well—and their operators were assuming that he was on or near the surface of the lithosphere.

The agent cast his memory back over the paths these things had been following, and decided that they might indeed be explained on the assumption they were seeking something and had a very restricted range of sensory perception. He dwelt for an instant on the last assumption, finding it unpleasant.

The radar beams, then, must have been used to track him. He had felt no such impulses, since digging in, although a portion of his hull remained exposed. But his attention had been so completely taken up with his work that he might not have noticed. He began to listen more carefully for electromagnetic radiation, and heard it immediately. On the instant, any doubts that might have remained concerning the intelligence of this race were disposed of.

There was a single source, which seemed to accompany only one of the machines, though the agent found it a little harder to locate precisely than the seismic sources. Apparently radio waves were being reflected from surfaces not in his mental picture of this part of the planet, thus confusing slightly his attempts at orientation. He was disturbed by the seeming fact that only one of these operators talked—and wondered why there had been no answer.

That problem was quickly solved, however. More careful listening disclosed a response coming from a fixed point some distance away. The agent did not attempt to make a seismic check on the environs of this source of radiation, since there was already enough to occupy his attention.

Still, why should only one of these machines, or its driver, be engaged in long-range conversation? Surely the others, if they were fit to be trusted to drive such devices, must occasionally have ideas of their own. It did not occur to him that the impulses might not represent speech—their pattern complexity was too great for anything else, though their tone was rather monotonous—quite literally. The frequency was constant and only the amplitude was modulated.

One possibility, of course, was that there was only one operator present, who was reporting to or discussing matters with his more distant fellow while he controlled all six of the nearby machines. In that case, however, the impulses he was using to control the subsidiary vehicles should be detectable, and nothing of the sort had reached the agent's senses.

Could it be that the orders were transmitted by metallic connections instead of radiation? They would have to be flexible, of course, since the relative positions of the machines were constantly changing! Yes, that could be it!

IV

Candace Parsons prepared dinner that night in the larger tent, over the fireless cooker. Because, for all of her native independence of spirit, she enjoyed being a woman and Hal's wife, and because she found herself not yet able, either intellectually or emotionally, to accept what had happened to them during the day, she concentrated on preparing the best meal possible under the circumstances.

While Hal and Truck continued to work the radio, under the trailer tarpaulin, she opened a couple of cans of chili, reinforced their contents with extra powder and placed on the stove a panful of the nourishing Mexican dish. She got a pot of coffee smoking, fried a dozen quarter-inch thick slices of bacon, and stirred the heated chili. Then she carried out to the men large, steaming platefuls of the rib-sticking food. They looked, she thought, as damp as she felt.

Returning with the coffee, she found the plates barely touched and told them, "You'd better eat hearty, characters. Heaven only knows when I'll cook another mess in this downpour."

Hal looked at her sheepishly, turned away from the transmitter and picked up his plate. "Sorry, baby," he said. "But, whatever this is all about, we seem to be it. We've had about everybody but the President on the air, asking us what in hell is going on."

"Let me take over for a while," she urged.

Truck, temporarily deserting his chore at the crank-battery to consume his victuals in what appeared to be three immense mouthfuls, said, "I'm afraid, memsahib, that this is going to be an all-night beat."

It turned out to be just that, since a baffled, excited, curious and somewhat frightened world refused to leave them alone. Increasingly, it became apparent that no other observer, human or electronic, had spotted further passage of the mysterious flying object which Truck called "the Greatest Whatisit." Thus, until the rain let up and the ceiling lifted, the Parsons Expedition was definitely roped down, and committed.

And the rain, as Candace had foreseen, did not let up. That, in itself, was one of the most unusual elements in the situation. According to all reports, the storm that had engulfed them extended over no more than a few square miles, centering upon the valley, and its surrounding territory.

The circumstance caused Truck to suggest, "Well, a limited storm area ought to simplify things. All they'll have to do now is locate the center of the cloud region. The center will be it—granted our Whatisit is seeding the clouds around here with malicious intent."

"How could it seed clouds that didn't even exist until after it came down?" queried Candace.

"So it made its own clouds," Truck suggested breezily. "Take it or leave it."

Candace and Hal Parsons exchanged a puzzled look. It was Parsons who got to his feet. They were sitting, Turkish fashion, on the ground beneath the larger tent's shelter and Parsons arose quickly to say, "Never mind the battery for a moment, Truck. Where's that Geiger counter of yours?"

"You mean to say you're gonna hunt uranium *now?*" Truck asked good-humoredly, as he complied.

"Not exactly," said Hal Parsons, a trifle grimly. "Okay, Truck—thanks." Neither he nor Candace felt up to putting into words the way they felt about the fear that gripped them. Any object capable of emitting radiations which could create such a furious local storm might well be capable of emitting radiations deadly to all human life within its radius. Both had visions of the Japanese fishermen who, in 1955, were caught in a radioactive fallout hundreds of miles from the Eniwetok atomic testing grounds.

Candace carried the electric lantern as Hal made his way about fifty feet downhill, into the saddle of the pass where they were encamped. She heard the ominous click-click-click of the counter, as her husband turned it on, and caught the strange, tense look on his features—a look sharpened by rainwater and the lantern's bright beam. She had an odd, shafting thought that this was not her husband at all, but a stranger. Quickly she closed mental lock and key on the idea, lest it be a prelude to panic.

He said, "It's high, but that might be caused by any number of reasons."

Candace nodded, and he took a few more strides. Then he bent low over a large puddle, formed in a hollow of the ground, and held the counter directly over it. The speed of the clicking increased by a clearly audible margin. After a moment, he stood up, and turned the instrument off.

"We're okay," he said. "The

rain is radioactive, all right. But, unless we hang around here for a month or two, it's not likely to cause any permanent damage."

"It can't damage *my* permanent," she replied. "I haven't had one in six months, and the coil came out in a week." The moment she had spoken, she felt like an idiot for making such a remark at such a time. On the other hand, she thought, this might be a moment when idiocy could really serve a purpose.

Hal said gently, "Shut up, baby," and then they walked back to the camp in silence, their minds full of oddly-parallel thoughts, hopes and fears. Increasingly, via radio and their own evidence, it was becoming clear that the Whatisit had elected to come to earth nearby, and probably knew exactly what it was doing.

"You know," said Hal, after lighting a damp cigarette, "if our friend had such a thought in mind, he couldn't have figured out a better way to stay clear of observation. Locating him on foot, in this rain, is going to be next to impossible. And the authorities outside the area won't find it easy to get in here. Our trail is washed out by this time. In fact, for all we know, the valley behind us will be flooded by morning. They can't observe from the air, because of the clouds—and the ceiling is so low I don't believe a helicopter could make a landing anywhere near here."

"Maybe he *does* want to," said Candace, relighting her cigarette thoughtfully.

"But that suggests . . ." Hal looked at her oddly.

"It suggests intelligence, all right," his wife said quickly. "And so did the swift, sure way he steered a path around this mountain yesterday. The big question now is—what *kind* of intelligence?"

"You're giving me the creeps," said Hal. He looked at her in the light of the electric lantern and smiled. But there was no mirth in his smile and when her hand crept toward his along the moist ground, he gripped it almost eagerly.

Truck MacLaurie stood over them. "If you two lovebirds are interested," he said, "I just got word they're sending a plane over in five minutes, to try to drop a flare through the clouds. They'll want to know if we can see it—and where."

The Parsons' scrambled to their feet and waited, by the radio, as the minutes ticked by. An eternity seemed to pass before they actually heard the distant drone of the plane. It grew rapidly louder and, all at once, appeared to be almost directly overhead. The receiver crackled, and Hal Parsons took over.

"You're coming in," he told them. "Parson here, *Over*."

"Roger dodger, Professor," came the buoyant voice of the airman overhead. "We're dropping a flare in five seconds. You should see her in twenty-six, when she blossoms. If you spot any little green men, let us know."

"Fire away," said Parsons. He frowned and added tersely: "And stop clowning."

"Roger dodger," was the reply, and Parsons wished, briefly, that the over-carefree birdman had to take the brunt on the ground with them. Then he recalled Candace's inane remark about her permanent, and it occurred to him that some people found such flipness an antidote to unendurable tension. He waited . . .

The flare burst, no more than half a mile away, its brilliance muted by the heavy mist and rainfall. Of the valley itself, it revealed almost nothing. Then, slowly, it burned out, leaving the darkness darker than before.

Parsons reported it, not too exactly under the circumstances, and the pilot said, "Well, that tells us exactly what we knew before. Stay with it, Professor."

Curiously, Parsons thought, he sounded discouraged.

Morning dawned, grey and soggy. But even so, the three on the mountain pass were lifted up in spirit by the renewal of light. The rain continued, without letup, and patches of mist clung to the slopes above and below them—and as far as their vision could penetrate.

They breakfasted on fresh coffee and the warmed-up remnants of the meal they had been unable to finish the night before.

"I never thought I'd have a miniature lake to wash dishes in," said Candace, dipping the plates in a puddle of fresh rainwater, and wiping them dry with a towel. "I've always had to scrub plates with sand on trips like this."

"Yeah," said Truck MacLaurie, "and radioactive rainwater, at that."

"Shut up, Truck," Hal Parsons said sharply, wishing he had held his tongue. It occurred to him, for the first time in his life, that people who can face grim reality and joke about it are, perhaps, far better realists than those who regard it so seriously that even talk of it disturbs them. What was troubling him was not the fact that the rain was mildly radioactive. It was the possibility that the great Whatisit might be emanating radiations of an alien nature, and more deadly to humans than anything the Geiger counter could pick up. Ignoring Candace's silent reproof, he walked slowly to the jeep.

Even though the slope into the valley was not steep, getting down the western side of the pass proved a far more difficult task than hauling the jeep up the east side had been. The reason, of course, was the unremitting rain, which was turning the poorly fastened dirt-and-sand hillside surface into treacherous, slippery rivers of silt and mud.

On this part of the trip Truck rose to heroic effort. Almost at the valley floor the little vehicle unexpectedly side-slipped into a freshly made brook, causing its rear wheels to stick and the trailer to fall over on its side. In a matter of seconds, the big football player had leapt from the rear seat of the jeep into the shin-deep muck, and was heaving at the trailer, with his neckcords swelling.

Before Hal or Candace could reach him he had unlocked the coupler and was hauling an upright trailer out of the water by main strength.

"The tarp held tight," he said cheerfully, not even panting. With his hair plastered over his forehead and his clothes clinging to him in the wet, he looked as if he had just stepped out of a shower with his clothes on. He added smilingly: "Get behind that wheel, Doc, while I push."

It took their combined efforts, but they finally got the jeep clear of the water and back on reasonably firm soil. Candace returned to the shelter of the jeep-top, while Hal and Truck recoupled the trailer.

Feeling thoroughly ashamed of himself for his previous sharpness, Hal said, "Truck, I'm sorry if I've been riding roughshod over you, but this whole business has me on edge. I mean, with Candace, and —" He let it hang.

Truck laid a massive, damp hand on Hal's already soaked shoulder and said with a grin, "Doc, don't worry about me. I've been chewed out by so many coaches giving me hell in the locker room that I don't mind a little ribbing from a guy I respect."

For some reason, the atmosphere lightened, though the rain continued to fall—and, curiously, the going grew easier from then on. Twenty minutes later, they had reached the floor of the valley, which extended almost level into the mist that blocked the mountains on the further side.

"Well," said Truck from the rear seat, as Hal slowly brought the jeep to a halt, "now that we're here, what do we do?"

The Parsons exchanged a look. Until then, reaching the valley had loomed as so large a problem in front of them that they had not considered the next move.

Candace laughed and said, "I'd pause at this point to powder my face if it would do any good in the dampness—if I had any powder handy."

To his considerable surprise, Hal found himself paraphrasing a long-forgotten and very ribald old Negro ditty which by rights should have remained buried in the rather scant excesses of his youth. He said, "It's right here for us, and if we don't find it, why it ain't no fault of its."

"Hey, Doc!" said Truck. "Where'd you pick that one up?"

"Probably," said Candace dryly, "in the very place where he picked me up."

It was the younger man who spoke seriously then. "No fooling, folks," Truck said. "Now that we're here, just how do we go about *finding* our inhuman friend? Don't forget—you're the brains in this pitch. I'm just the muscle."

"I'm afraid we're going to have to use that Geiger counter of yours again," Hal Parsons said. "And no cracks, please! If we can find any variation of intensity in the rainfall radiation, there may be a chance. . . ."

"Gotcha, Doc." Again, the young Goliath was out of the jeep, and working at the trailer tarpaulin.

"Do you think it will work, honey?" Candace asked.

Hal Parsons shrugged. "It might," he said. "It just might."

But it didn't. As remorselessly as the rain continued to fall, the mild radioactivity continued to register without variation. After testing puddles for two hours, the two men returned to the jeep, where Candace had coffee ready for them once more. She asked no questions as to the success of their experiment. One look at their faces as they emerged from the mist told her all she needed to know.

"Carnotite," said her husband, lifting his face from an empty tin cup and wiping his mouth on his sleeve. "Not enough to report on —just enough to bitch us up for an hour, with a false lead. You might not believe it, but this is one hell of a big valley."

"It keeps getting bigger," said Truck MacLaurie mournfully, through a coffee mustache. He looked at Hal Parsons and asked, "Well, Doc—what next?"

Parsons was trying to come up with some sort of a constructive reply, when Candace motioned for him to be silent and lifted her face upward. The others followed her gaze and saw nothing but clouds, rain and fog. Then they heard it —the drone of a plane directly overhead. Without a word, both men handed Candace their empty cups and moved toward the trailer.

It was a mere matter of minutes, before they had the radio back in action, and were trying to communicate with the crew overhead. While Truck cranked away at the battery, to raise power, Parsons hung onto the transmitter, urgently repeating, "Parsons calling plane. Parsons calling plane. We hear you. We hear you. Come on in. Come on in. Over . . ."

All he could get, on the earphones, was a rumble of static, through which, now and again, he heard the faint, unintelligible mutter of the operator upstairs, trying to break through. Candace looked at him anxiously, her hair oddly slicked into bangs by the rain. He shook his head hopelessly.

"Keep trying," she said softly. "Keep trying, Hal honey."

Frustration was high within him, but he nodded and tried again. "Parsons calling plane. Parsons calling plane," he began.

This time, there was no doubt about the answer. It came, clear as a voice in some unbuilt next room, saying, "Parsons calling plane. Parsons calling plane."

"Who's that?" he barked, recalling the impertinence of the aircraft radio message of the night before.

"Who's that?" he barked. *"Hello?"*

"Who's that? *Hello?"* the mocking voice replied.

Parsons mopped rainwater out of his eyes and snapped, "What in hell is going on? This isn't funny, Mack!"

And the voice replied, "What in hell is going on? This isn't funny, Mack!"

"Cut it out, you joker!" he said furiously. "If you've got a message for us, unload it and take off."

To which his tormentor retorted, "Who's that? *Hello?"*

"Hal honey!" interposed Candace, who had crowded close and turned back one of the earphones to catch the mocking message. "Hal honey, he's replying in *your* voice."

"So what?" her husband countered. "Whoever he is, I'm going to see he gets hell—once we're out of here."

"Just a second." She nudged him clear of the transmitter, bent over the mouthpiece, and said clearly, "Toodle-oo, old thing."

The answer came back clear as spoken—and in perfect reproduction of Candace's voice. "Toodle-oo, old thing."

They stared at each other until Truck came over. He pushed back his hair and said, "What is this—a private game, or can anybody play?"

"It's beginning to look," said Hal quietly, his controlled voice belieing the wild excitement in his eyes, "as if your great Whatisit is as anxious to get in touch with us as we are with him." Then, turning to Candace, he asked, "What do you think, baby?"

"I think," she said, "if I were an alien and wanted to be a radio announcer and could only receive H. V. Kaltenborn, I'd give it back to him just the way he was giving it to me."

V

IT BECAME INCREASINGLY evident to the Conservationist that he could lie there, until he was trapped in an earthquake, making up five hundred theories per second, without getting one whit closer to knowledge of what was happening around him. He was going to have to examine the machines more closely. The only question was one of tactics. Should he go to them, or have them come to him?

He decided first to try the second gambit, since it offered more promise of drawing out information as to their nature and abilities. He would thus be able to determine precisely what stimuli affected their senses of equipment, and the extent of their capability in analyzing what they did detect.

Naturally, not a wave of their radiation had, thus far, conveyed any meaning to the Conservationist. More accurately, the few patterns that even remotely matched patterns of his own language did not deceive him for an instant by such chance similarities. Nor did he suppose the natives would have any better luck with his language.

His first attempt at attracting their attention consisted merely of broadcasting sustained notes on a variety of frequencies, other than the one they were using. As he had rather expected, these produced no noticeable reaction. Travel and conversation went on unaffected. When he repeated the attempts, using the same wavelength as the natives, however, the results were just as unsatisfactory. It was extremely frustrating.

Travel stopped, and after he had repeated the signal a few times, all six of the vehicles seemed to come together at one spot. In the pauses between his own transmissions, the native speech sounded almost continuously. Yet he felt doubt that he had even been heard.

He had rather expected that there might be an attempt to respond to him in kind, but this did not occur, even though he tried sending out his wave in various long and short pulses which should have been easy to copy. At least, he used lengths corresponding to those of the radar pulses which he had felt at his arrival, and which had, presumably, been emitted by members of this race.

They failed to respond to the patterns, however, even when in desperation he increased the lengths of the bursts of radiation to three or four thousand microseconds. The very speech patterns of the natives changed carrier amplitude in shorter periods than that—they must, he felt, be able to distinguish such intervals!

The agent began to speculate upon the general intelligence-level of this alien new race. He had to remind himself forcibly that, since they could move around so rapidly, they must be able to design and build complex machines. It was startling, to say the least.

Then it occurred to him that *all* the vehicles he was watching might be remote controlled, that the electromagnetic waves he was receiving were the control impulses. Yes, yes, that must be it! He spent some time, trying to correlate the radio signals with the motions of the machines. The attempt, of course, failed completely, since men are at least as likely to talk while standing still, as while walking around.

This proving a poor check on his hypothesis—it did not disprove it, since the machines might be able to do many things besides move around—he tried duplicating some of their complete signal groups, watching carefully to see whether any motion of the vehicles resulted. He realized that the controlling entity might not like what he was doing, but he was sure that satisfactory explanations could be made, once contact was established.

The result of the experiment was a complete stoppage of motion, as nearly as he could tell. It was not quite what he had expected. But there was some gratification in getting any result at all. For several whole seconds there was silence, both seismic and electromagnetic.

Then the native speech—it had to be speech—began again, in groups which still seemed long to the agent, but which were certainly much shorter than most of those used before. He duplicated each group as it came.

"Who's that? *Hello?*"
"Who's that? *Hello?*"

"What in hell is going on? This isn't funny, Mack!"
"What's going on? This isn't funny, Mack."
"Cut it out, you joker! If you've got a message for us, unload it and take off!"
"Who's that? Hello?" The agent decided the last signal group was too long to be worth imitation, so he went back to one of the earlier groups. This action resulted in brief silence, followed by a pattern, brief, but with a fresh modulation, which he mimicked accurately. For several whole minutes, the conversation, if it could be called that, went on. He felt real pride now, a self-congratulatory kind of exaltation in being able to carry off his cleverly assumed masquerade with perfect confidence, vigor and, certainly, no small measure of success.

The Conservation agent had decided long since what the native machines would almost certainly do, and was pleased to detect them getting into motion once more. But when they had gone far enough for him to determine their direction of travel, he discovered, with some disappointment, that they were not moving toward him.

He would have had little trouble solving their motives, had they been moving straight *away* from him. But the angle they took carried them more or less in his direction, albeit considerably to one side. He found this a complete mystery, at first. Finally he noticed that the group was traveling along a depressed portion of the lithosphere's surface, and seized upon, as a working hypothesis, the idea that their machines found it difficult, or impossible, to climb slopes of more than a few degrees.

In that case, of course, they might not be able to reach him, directly or otherwise, since he had buried himself some distance up the side of a valley. He considered again leaving his position and coming to meet them, but reached the same decision as before—that he could learn more by seeing what they did on their own.

They spoke rarely as they traveled—but the agent found that he could always make them broadcast, by ceasing to radiate his own signal. Had they not been pursuing such an odd course, he would have supposed, from that fact, that they were using his radiation to lead them to him. His radiation! However, they kept on their course until they were somewhat past its nearest point to his position before they paused. Then there was a brief interchange of signals with some distant native, apparently in an atmosphere machine, and travel was resumed, at right angles to the original direction.

Now, however, the vehicles were heading away from the buried ship, had, in fact, turned left. The Conservationist gave up theorizing for the moment and contented himself with observing. He repressed his mounting excitement and became as still as a figure of stone.

They did not travel very far in the new direction. In less than half an hour they stopped again, held another brief conversation, and then began to retrace their steps to, and finally across, their original route. Apparently, they were still interested in the agent's broadcasts. At any rate, they continued repeating the early "Hello" and "Who's that" signals to which he had originally responded, whenever he stopped radiating. They were not following the radiation, but certainly—almost certainly—they had some interest in it.

Then, quite abruptly, they stopped traveling and appeared to lose interest in the whole matter. The group broke up, and its members wandered erratically about for some time. Then they drew together once more and gradually quieted down completely, or at least to the point where the agent could not be sure that the occasional impulses coming from that area were due to their motion.

He had just developed another theory, and this new trick bothered him seriously. He would have preferred to ignore it, but he could not. It had occurred to him that these creatures might be able to detect electromagnetic radiation of the sort he had been broadcasting, but not be able to identify the direction from which it came. He had heard of cases of physical injury among his own people which had produced such a result.

The idea that such a disability might be universal in this race called for a severe stretch of the agent's imagination, but he toyed with it all the same.·As a result, he had just come to realize that the peculiar motions of the things he had been observing could indeed be accounted for by the assumption that they were searching for him under some such handicap—when they stopped moving. This was hard to reconcile with any sort of search procedure. What possible reason could stop them? He wished sometimes there could be fewer complexities in his existence. What possible reason?

Lack of fuel? Inconceivable, assuming even minimum intelligence on the part of the operator or operators.

Surface impossible for the machines to travel over? Unlikely, since several of them had come some distance toward him during their erratic wandering after the halt of the main body. And there had been others in the atmosphere.

Sun-powered mechanisms, halted by the fact that night had fallen? It was possible, though it seemed a trifle odd for such a device to be used on a rotating planet, where it must be sunless half the time. Also, it seemed doubtful that the machines were large enough to intercept the requisite amount of solar radiation. The agent had a fair idea of their size and mass, from the minimum observed separation, plus the energy with which they struck the ground.

Not interested in him at all, and stopped simply because they had reached their intended destination? This seemed all too painfully probable, if the course of their travels were considered by itself—yet nearly impossible, if their reaction to his broadcasting were taken into account.

It was at this point that the agent began to consider seriously the possibility that he might never be able to get the information of their danger across to the inhabitants of this planet. Their behavior, so far, seemed to lack any element he could recognize as common sense. He was open-minded enough to realize that this might work both ways, yet such a possibility did not augur well for the chances of successful communication between the two intelligences involved. There were cynics even among his own people who claimed that folly and ignorance always went arm in arm, and were biological constants throughout space.

Once more, he was facing the question of whether he should go to meet these gadgets, or wait where he was—and, in the latter case, how long he should wait. Certainly, if he were to check the possibility that they were sun-powered, he should not stir until after night was over.

But none of the other hypotheses could very well be tested without actually examining, at close hand, the natives and their machines. He decided, then, to wait until sunrise, and for a reasonable period thereafter. Then, if these things did not resume their journey in his general direction, he would seek them out.

As it turned out, he did not have to move. The appearance of the sun saw the vehicles already in motion, which was informative in a negative way. After a brief period of random traveling, they congregated once more, seemed to confer silently for a time, and then resumed travel along their former route. Also, they broadcast once more the signal the agent had come to interpret as a request for him to start transmitting.

The events of the preceding afternoon were repeated in some detail. The group continued past the agent's station on their straight-line course for a short distance, then stopped, and once more made a right-angle turn. This time, it was to the right, toward the hidden alien—and the agent realized that this theory about their sensory limitations must be at least partly correct.

They had to go through elaborate maneuvers to locate the source of a radio broadcast—maneuvers which suggested that even their ability to judge the intensity of the radiation was rather crude. It took them about a tenth of the planet's rotation period, this time, to narrow the field down as far as their radio senses appeared to permit.

Before mid-morning they had made two more right-angle turns, and then spread out to cover, individually, the remaining area of uncertainty. The agent settled comfortably in his hole and awaited discovery. This should tell him much.

Just how close would these things have to come to detect him directly? Would he be able to pick up their nerve-currents first? What would they do when they found him? How long would it take them to realize that he was not a native of their world? And, most important, would they have some constructive ideas about means of communication? Who did he think he was fooling? At the moment the agent would have admitted to anyone that he himself had none. And if he was up against a blank wall in that respect, how could he reasonably expect them to come up with something really new and brilliant?

He kept his own senses keyed up, striving to detect the first clue, other than radio and seismic waves, of the nearness of the Earthly machines. Presumably, they were more or less electrical in nature, and he knew that electric and magnetic fields must, sooner or later, draw close enough to give him a picture of their structure. A little closer than that, and the electric fields of the operators' nervous systems should permit him to deduce their shapes and structures—assuming, of course, that at least one operator was with the present group of machines, which could hardly yet be considered certain.

Although it was the machine with the radio that actually stumbled on the buried vessel, the radio was not in use at the time. As a result, the agent decided, rather quickly, that no operator was in fact present. The radio was, of course, put to use the moment the ship was sighted—but its structure and nature was obvious to the alien, and it was quite evidently not an intelligent being.

It was, however, the only object in the vicinity with functioning, electrical circuits. Moreover, there was no direct sign of life in any of the machines which gathered quickly around the ship. Finding it a little hard to believe even his own theories, the agent once more examined the radio—only to reach the same conclusion.

Its organization was not sufficiently complex to compare with a single living crystal, much less an entire nervous system. The conclusion seemed inescapable. Not only was the machine carrying it being controlled from a distance, but even the vehicle itself operated *without detectable electrical forces.*

The machine, of course, could not be invisible. His failure to see it meant merely that he was employing the wrong means—*anything* material can be seen, in some way or other. There remained the question of just what *were* the proper means in this particular case.

Free metals affected electric or magnetic fields, or both, in ways which permitted their recognition. Only a few fragments of such material were present—fragments quite evidently shaped by intelligence, but not themselves part of either an intelligent body, or even a complex mechanism.

Non-conducting crystals reflected and refracted many kinds of radiation. Perhaps these things, then, could be *seen.* The only trouble with this idea was that eyes were not a normal part of the agent's physical makeup. While his ship possessed several which were used in navigation—stars were most easily detected and recognized by light waves—they all happened to be underground at the moment. He had never anticipated a use for them on the surface of the planet, not being himself a chemist.

The machines were now all moving about on the ground in his immediate vicinity. One of them even moved onto the exposed section of his hull for a few moments and it gave him his first chance to approximate their mass really accurately. Unfortunately he could not determine precisely how much of the energy radiating from their footsteps was due to weight.

The machine on his hull carried a tiny ionization tube, whose behavior at the moment was being affected by the mild radioactivity of the ship—activity only natural after a million years in interstellar space. The purpose of the tube was no more obvious than that of the electromagnetic radiator. Neither could move or think. The only possibility seemed to lie in a connection with the remote control of these machines. Perhaps, they were sensing devices of some sort.

There seemed no logical reason for not raising the ship far enough to get a look at these alien machines. He had discovered all he could expect to learn, from where he was. They *did* receive him. They *were* interested, and they, therefore, had at least glimmerings of intelligence. They could *not*—or, at least, their machines could not—determine the direction from which radio-waves were coming.

It was still not clear to him whether these machines were under the control of one individual, or that of several. There seemed no way of investigating this important question for some time to come. What the agent wanted to know, as soon as possible, was just what sort of mechanism could operate *without perceptible electrical fields*—and that seemed to demand that he *see* them. Yes, he must see them.

His hull had long since cooled, and could be controlled without difficulty. He started it vibrating again, and, simultaneously, applied enough drive to counteract the weight of ship and its contents. For a fleeting instant, he wondered whether the distant operators could detect the flickering of the myriads of relays that responded to his thoughts, or even the electrical fields of the thoughts themselves.

If the latter were true, they could certainly not interpret them properly. In that case, the machines would have found him much earlier, and the agent would, by now, have been holding a conference with them about the best means of intercepting the mole robots. That possibility, he decided, could be ignored.

The patrol flier lifted easily, until over half its bulk was above the ground. Its pilot held it there, briefly, while the rhythm of the hull packed and firmed the powdered soil that had drifted beneath it. Then he cut his power once more, and began to look about him with his newly uncovered eyes.

VI

THE LITTLE PARTY'S jubilation had proved short-lived. They had, it was true, attained communication with the Whatisit—but apparently all that it could or would do in this field was to mimic their voices and speech in startlingly unexpected fashion. After a quarter of an hour of ever-increasing exasperation, Truck MacLaurie won Parsons' temporary disfavor by suggesting, "Hey! I wonder if it can sing."

Candace didn't help the geologist's feelings by laughing outright at the infantile remark.

Hal said, "It's not funny, dammit! How are we going to get any more sense out of it. You'd think from the way you act this was a Sunday School picnic—not something deathly serious, even terrifying."

"I guess we'll have to find it first," said Truck, rubbing his face briefly dry with a large blue bandana. He looked more troubled and uneasy than he had permitted himself to look a moment before.

Candace gazed sadly at the ruin of her cigarette. "I wonder," she said, "just how we're going to accomplish that."

"Follow the beam," Truck suggested. He spoke lightly, but all the levity was gone from his stare.

"Get in," said Parsons, nodding toward the jeep. "We're going to find out if our friend really *is* beaming his messages."

They drove a quarter of a mile and tested. The baffling mimicry aped them just as clearly, just as strongly, as before.

"Maybe," said Candace optimistically, "we're headed straight toward him."

"Not likely, dear," said Parsons. But he got the jeep going over the rough terrain at right angles to their previous direction before making another test. Once again, the mockery continued without any noticeable fluctuation in volume or alteration in its monotony of tone.

"*Damn!*" he exclaimed fervently. "He's sending without direction."

"What makes you so sure it's a *him?*" asked Candace.

"All right," said Hal, a trifle testily. "*She's* sending without direction."

"I didn't mean it that way, Hal honey," Candace told him. "I was just wondering if we hadn't jumped the gun in thinking of our friend as an intelligent entity."

"He, she or it was smart enough to move around that mountain yesterday," put in Truck, from the rear seat. "That took brains."

"Or machinery," said Candace.

"Supposing its nothing more than a *machine*."

"That," said Hal, resting his forearms on the wheel in front of him, "raises some mighty interesting possibilities. Let's say, for the moment, that it *is* a machine. Obviously, a missile—if that's what it is—could have reactors that would enable it to avoid a crash—as with the mountain. But if it is a machine, somebody, or some *things,* had to make it. No intelligent creature would manufacture anything so complex without a purpose, and send it at random through space."

"Maybe it's not from space. Maybe the Commies sent it over to broadcast germs or something," said Truck.

"You think of the loveliest ideas," said Candace. Then, frowning and poking at the sopping ruin of her hair, "If that were true, it wouldn't be answering us—even with mockery. It would be lying nice and doggo. My money—listen to the girl!—says it's from space. If it were a missile that goofed, you can be sure the big brains in the Pentagon wouldn't be kicking up such a fuss."

"Well, we aren't going to solve the problem by sitting here talking about it," Hal said practically. "We've got to hunt until we find it."

"How are you going to do that?" Candace asked.

He told them. They were going to do it on foot, tracking the valley floor and leaving bits of cloth and direction markers whenever they reached the hills, so they would not be forced to retrace their steps. "That way," he concluded, "we can find out where it isn't, if nothing else."

"We can get good and wet, too," said Truck.

Parsons quelled him with a look, and they got busy. They hardly spoke at all, for their thoughts were now completely immediate, grim and serious.

It was a tedious, unrewarding day of plodding through rain-soaked sand and soil. When, as the sunless daylight waned, they finally returned to the shelter of the jeep, all three of them were exhausted.

"Another two or three days of this," Truck complained, "and my legs will be too musclebound for football." It wasn't what he'd intended to say. It was merely a quick cover-up to conceal his real emotions.

"I think I left my feet on the other side of the valley, last time across," said Candace, falling in with his mood. "Hal honey, where do you suppose it is?"

"It's here somewhere," said Parsons, wishing his own feet would cool off and stop aching. "We just haven't looked in the right places."

"We'd better get back up a hill and do some broadcasting," said

Candace. "I'll cook us some sort of a meal."

"I'm too tired to eat now," Parsons told her. "But you're right." He got the jeep into gear, adding, "Maybe they've found it somewhere else."

"Happy thought!" said Candace. "But it's too much to hope for."

And theirs *was* the only report on the alien. Parsons talked to a General Somebody, who had jetted from Washington, D. C., to Butte that afternoon, to be closer to the critical scene. Apparently, the entire world was in a ferment over the possibility of contact with a messenger from an alien race.

"How are ground conditions?" the general asked.

"Lousy!" Parsons told him bluntly. He gave him a succinct account of the frustrating day the expedition had endured.

"You mean, you actually *talked* with it?" the General asked.

"You could call it that," said Hal, and went into a full explanation.

"Do you think we could get a helicopter in under those blankety-blank clouds?" the General wanted to know. "It would enable us to get a fix on its whereabouts."

Parsons looked dismally at the mist that enshrouded hilltop and valley alike. "Not a chance, I'm afraid," he said. "This stuff is thick and close. We're snafu-ed, *but good!*"

Candace, who was standing by with a plate of hot food, heard this portion of the conversation and said, "Hal honey, maybe if they could get a plane overhead and they knew where we were, we could rig some sort of a fix on our friend. Ask him?"

"The trouble with that," said Parsons, "is our pal's sending doesn't reach up here. And how are we going to tell where either of us is if we can't see through the clouds?"

"What's that?" the General asked. "What's going on?"

"Mrs. Parsons," said Hal. "She wonders if you couldn't send a plane over tomorrow to help us get a radio fix on our friend."

There was silence. Then, "Tell your wife she gets a large box of filter-tip cigars when this is over. By God! That's the first really constructive idea that's come out of this foulup yet. But it will take a bit of doing. Lucky that stuff over you is not much more than two thousand feet. You'll hear from me in an hour. Signing off and good luck."

"What did he say?" Candace asked eagerly, as Parsons flipped the switch and motioned for Truck to stop cranking the battery.

"He says he's going to give you a box of choice Havana cigars when we get out of this hole, baby," Parsons told her, accepting his food. "Mmm! These beans are good! What did you do to them?"

"Oh—I just let you work up an

appetite, that's all," said Candace. Then her eyes widened. "You mean he's actually going to do it?"

"He's going to *try,*" said Hal through a full mouth. He tilted his tin plate to let the rainwater trickle off onto the ground. "If we ever do make sense with this creature, I'm going to ask him to turn off the waterworks."

"Amen to that!" said Candace.

"I was figuring on working up a sunburn that would last all winter," said Truck mournfully.

The general radioed back, on the nose. An air-fix would be attempted the following morning at ten o'clock. It was complicated, but he thought it could be done. "We've got to find that thing—or rather, you have to find it. Are you aware that we have an expedition with Weasels on its way to reinforce you?"

"Weasels!" Parsons was startled. "But we got in here okay in a jeep."

"You couldn't do it now," the general told him. "Those two days of wet weather have washed out all the trails. But don't worry. We'll be getting through to you soon. Just find our friend and see that he doesn't take off before we open communications."

"What's the verdict to date?" Parsons asked. *"Is* it extra-terrestrial?"

"Looks that way. The Russians swear on a stack of Karl Marx *they* had nothing to do with it. They're talking it up as some new sort of war-mongering frightfulness we've developed. Well, I'll be overhead tomorrow morning."

Once again, there was little sleep in the expedition. But their restlessness was not the result of frustration, unrewarding as their day of effort to locate the stranger had been. There was a sense of impending excitement, of discovery lying just ahead of them, a growing awareness of the importance of the position fate had put them in.

"If they're right," Candace mused aloud, "you and I, honey, are the first two humans ever to communicate with a being from another world."

"What price communication?" said Hal. "We might as well have been yelling our heads off in Echo Canyon."

"How about me?" put in Truck. "Don't I get to talk to it, too?"

"Of course, Truck!" Candace said warmly, reacting with quick, feminine sympathy to the young gladiator's sense of having been left on the outside. "You can talk your varsity team mask off tomorrow."

"Gee—thanks, memsahib," said Truck, feeling his dark inner mood lighten a little.

He retired into silence, apparently considering the effect of his impending importance on certain members of Candace's sex. She and Hal exchanged meaningful glances. They were both growing increasingly fond of Truck. He might not be cut out for a Ph.D., but his strength and stamina, his amiability and his quick native intelligence made him a valuable member of the closely-knit team they had become.

With the coming of the dawn, they rose and broke camp again. They made another descent to the valley floor, handling jeep and trailer with extra care lest an accident damage their radio gear. Certainly, weatherwise, the situation had not improved overnight. Mist and rain were equally heavy, and the once hard-packed ground was slowly turning into a quagmire. It took them more than an hour to get located on a bit of high ground, where they would not become hopelessly bogged down.

"Let's see if our friend is still sending," said Candace.

They set up shop, and Truck took over the mike. He said, "Hello, out there," and promptly received a "Hello, out there," in response.

Parsons scowled at the set. "If our pal doesn't shut up when the General starts sending," he said, "it's going to be awfully confused."

"We'll manage," Candace said confidently.

The general, as usual, was on time. He said, "I'm somewhere overhead in a helicopter, with another copter standing by. We want a fix on you, first. Then we'll try for a fix on the alien and at least give you direction."

"Hello out there," said the voice from the stars.

"Who in hell is *that?*" the general asked, startled.

"That," said Parsons, "is our unexplained visitor. You'll have to sift if he keeps cutting in."

"Okay, Parsons—let's get busy," said the general. "Start reeling off a page of statistics—or anything that comes into your head."

Parsons complied with the multiplication table. After imitating him at first, the Whatisit apparently gave up and stopped sending. Ten minutes later, the general's voice came over the receiver.

"We've got you," he said. "Now, see if you can get the owner of that voice."

Parsons raised the unknown visitor, using short, varied sentences. He was, he felt with growing excitement, beginning to learn a little about the alien. Two or three times, when the human speeches were long and intricate, or merely repetitious, it had simply ceased sending. Evidently, some sort of selective mind was at work, determining which phrases merited repetition, and which did not—even though, apparently, none of them made sense to the alien.

"Okay, Parsons, here it is!" said the general. "Got a compass handy?" He gave the directions concisely, and concluded by saying,

"Sorry we can't give you more. We spot you maybe half a mile apart, but our own location is too unstable to give you a clean estimate of distance. If you follow the direction I just gave you, and keep your eyes open, you ought to find him."

"We'll do our best, General," said Hal. Then, sighting along the direction-line he had just been given, he exclaimed in dismay, "Damn! This runs right along the hills on the north side of this bowl."

"You'll manage," said the general, with a confidence Hal, at that moment, was far from feeling. "Good luck. But be careful. He may be dangerous."

"*Now* he tells us!" said Candace, who had appropriated one of the earphones.

They had to leave the jeep where it was, and scramble, slipping, stumbling, peering vainly through the mist for some sign of the alien. Their progress was abruptly halted when they had covered about a quarter of a mile, and the hillside across which they were moving became split by a sharp declivity.

"It couldn't be worse!" muttered Parsons. "We'll have to work our way around it."

Working their way around took them approximately half an hour. They were about halfway up the gentler slope of the far side when Truck, who had lumbered on ahead, let out a yell that echoed from crag to crag like a many-throated summons to battle.

"Here it is! I've found it! Come on, you two! *I've found it!*"

It was big. Although, in some unexplained manner, it had buried itself in the hillside, so that only a small sector of its top-surface showed above ground, the curve of its dull-grey, irregular and knotty metallic surface revealed a diameter of more than twenty feet. There it sat, immobile, apparently harmless—like a large piece of leaden-hued pewter discarded from a New England farmhouse attic.

"We found it! We *fou-ound* it!" Truck chanted, and then suddenly turned deathly pale, as the terrifying significance of the find's brooding stillness and nearness and alienness was borne in upon him.

"Get off at once!" Parsons almost shrieked the words. "You don't know what sort of radiations may be coming from it."

Truck swayed as if in mortal terror and scrambled down. "My God!" he breathed. "I left the Geiger counter back in the trailer."

"Get it," Parsons shot back to him. *"And get the jeep as close as you can."* Apparently whatever is inside that thing can only communicate through the radio."

"I know—sure," said Truck. "I —I'll be right back with the Geiger."

He had just turned to carry out the order, when Candace uttered a small shrill scream and cried, "Look! Hal, stay back! It's boiling the earth around it!"

Something very strange was happening. Invisible currents were making the once-sandy soil in which the object had settled seethe like boiling water in a kettle. As Parsons pulled his wife quickly away from the area of disturbance, he thought that her use of the word "boiling" had been singularly apt. From a safe distance up the hillside, the three of them watched the ground around the visitor act as solid matter was not supposed to act.

It was Truck who first sensed the visitor's intentions. Stabbing a large grimy forefinger at it he announced, "For Pete's sake, he's coming up!"

They looked on in awe as the dull-grey globe that was not from Earth slowly emerged from its bed of soil, looming larger and larger as it rose, and revealing in what appeared to be its nose a pair of opaque, circular objects that looked like eyes.

VII

THE STAR-TRAVELER already knew, of course, that he was in a valley, partway up one of the sides. The hills bounding it were not particularly high, especially by the standards of this planet. In fact, the Conservationist had a pretty accurate idea of the dimensions of the Himalayas, distant as they were—though he had been more interested in determining the rate at which they were rising. He gave the local elevations only a passing thought, then sought to examine what lay closer to his vision-outlets—outlets which the Parsons group had quite correctly labeled "eyes."

He failed. The details five miles away were clear and clouds of what must be water or ammonia droplets hanging at still greater distances in the atmosphere were still clearer. But, as he brought his attention to objects nearer and nearer to his ship, they grew, shapeless, and increasingly harder to examine.

Cursing himself for forgetting, he recognized the reason. His eyes were perfectly good instruments—for the purpose toward which they had been designed. They were carefully shaped lenses of calcium fluoride, designed with almost a full hemisphere of field and their curved focal surface was followed faithfully by the photosensitive material of his own flesh. The tiny metallic crystals in his stony tissues would, of course, be affected electrically by light, and, like many of his race, he had learned to interpret the light-images formed by lenses.

There was just one catch. There was no provision for changing either the shape or the position of the lenses. But actually, why should there be? They were designed to enable him to determine the directions of the stars, whose distances were for all practical purposes always infinite. He had never needed focusing arrangements until now.

The eyes were a foot across and almost as great in focal length. Objects a hundred yards away were blurs. At six feet they were scarcely interruptions to the background. He could just tell, by sight, that there were moving objects in his vicinity, and get a vague idea of their size. Beyond that, details were indistinguishable.

The nearest repair-shop where his machinery could be modified was about six thousand light-years toward the galactic center. He could, of course, pull his flesh back from one or more of the lenses until the eye involved focused at a distance of a few feet—if the situation would wait for the necessary years or centuries. However, even if the situation did wait, the natives and their machines probably wouldn't.

He could wait until they departed, and examine them when they were far enough away. Better than this, he could fly to a distance at which they were reasonably distinct in his sight. The question raised in that connection was, of course, how the natives would react to such a move on his part. However, if he did not move, he would probably learn nothing. Therefore, he resumed his rise from the soil, cleared its surface, and hurled his vessel half a mile upward.

To observe, and, in effect, to photograph the details of what lay below took only a few microseconds. Then he moved a few hundred yards to one side and repeated the procedure. Three seconds after takeoff, he was settling back into his original location with a fairly clear picture of the strange equipment surrounding it firmly painted in his mind.

He understood now why the seismic impulses had come in pairs. Each of the machines was supported by two struts, which were so hinged as to permit several degrees of freedom of motion. During his brief period of observation, they had traveled enough—away from the point where his ship had been resting —to permit him to analyze their startling method of travel. This seemed to consist in balancing on one strut, falling in the desired direction, and catching one's mass with the other before collapsing completely. The process was repeated cyclically.

It appeared, mathematically, that the value of the planet's gravitational acceleration would put an upper limit on the rate of travel possible by this means. The agent found himself a little dubious about the engineering advantages of it. If one had to travel on the surface, wheels seemed easier—although an irregular surface might present further difficulties. Few Conservationists, surely, had confronted problems so difficult to resolve.

At least, he had eliminated the last possible doubt that the things were non-metallic, non-electric machines, since he had actually seen them move in a manner which verified and complemented his seismic observations. This implied that the natives were not merely cultured, *but had developed a physical science equal to, perhaps greater, than that of the agent's own race.* The latter was certainly possible, since he had not the faintest idea of what was the operative principle of the devices. It was a disturbing speculation, but he refused to enlarge upon it emotionally. Oviously they had *some* electrical equipment. The signal detector and broadcasting device, as well as the ionization cylinder, were quite evidently as artificial as his own ship. Their science, regardless of its development, could not be entirely alien. It might be possible for him to learn something about it. If so, it was important that he begin—for the equipment needed to stop the moles would have to be obtained from these people in rather short order.

The agent examined once more, as precisely as his sensory equipment permitted, every detail of the things around him, which were now returning slowly, after their hasty withdrawal. He broadcast his *"Hello"* again, and carefully noted the way it affected the receiver. When the answer came, he checked with equal care the source of the modulating energy.

The result was interesting. The receiver apparently did not consider the carrier waves important. It damped them out and used, through most of its circuitry, a secondary signal consisting of the original modulations. This was caused to vary the strength of a magnetic field which, as nearly as the agent could tell, was used to impart mechanical motion to an object principally non-metallic.

He could get only a rough idea of its size and shape from the space left for it in the mechanism. The evidence seemed to indicate that the whole device simply rebroadcast the modulation of the original signal mechanically into the atmosphere.

He knew, of course, that a gas *could* carry compression waves, though it had never occurred to him that they might be of any particular use. He had simply never stopped to wonder why his method of digging was more effective on a planet with atmosphere. It did no good to blame oneself for such oversights when the fat was in the fire. Anyway, he was sure of one thing. The waves were being used to carry the signals controlling the machines. Certainly no others were. They also served for communication, since similar waves appeared to be received by the same disc in the signal device, and were used to modulate its broadcast electromagnetic impulses. This process seemed pointless, except as a means of long-distance communication. Probably pressure waves did not transmit energy so effectively through a gas as electromagnetic radiation carried it through space. So far, so good.

It all tied in, more or less, with the evident fact that these machines were not electrical, even if it did not begin to explain how they actually worked. Some sort of more precise analysis would, of course, be needed. The metal he could detect about the things seemed quite purposeless, and he did not see that it was likely to help.

It was present in small, disconnected bits and was devoid of electrical energy, if you brushed aside the minute currents generated by its motion in the planet's magnetic field.

The machines, then, were made virtually entirely of nonconductors, and should be about as easy for the agent to examine as a device consisting exclusively of gas jets and magnetic fields would be for a human being.

This meant that the analysis would have to be by highly indirect methods. A chemist, with his laboratory machine, might be able to do the job in microseconds. But a traveling device, like the scoutship, had no equipment designed with any such purpose in mind.

He suspected that this was one of the situations where the sensile members of his race—the great majority—would leap at the chance to show their superiority over one who was bound to a machine. It had always been that way. It was a common enough feeling among those whose lives were primarily intellectual. The doers, like the agent, countered it with a clear recognition of the necessity for their work. At the moment, however, the agent rather wished that a normal person had been present, to show his intellectual superiority.

Then he realized that his own possession of machinery did not disqualify him as an intelligent being. If a member of his race could solve this problem, it was as likely to be himself as anyone else. He would have to use *all* his knowledge, of course, not just the specialized information which was all the millennia of flight demanded.

Enough knowledge should be there. He had, of course, been

young when he had elected this life, but he had had much thinking time before his career was actually begun. Also, there had been a good deal of time to think as he drifted among the stars, and opportunities to gather data that planetbound thinkers had never possessed.

He would have to go back to the most elemental principles of thought—if he could. First, he had decided, on the basis of what seemed adequate evidence, that the planet was inhabited—that its inhabitants used machines and, therefore, had freedom of motion —and that these machines were based on a technology almost, but not quite wholly, alien to his own.

Nevertheless, the devices must operate under the same physical laws that obtained elsewhere in the universe. This meant that they must take in some form of energy, must perform a desired action, and must eventually account for the energy as heat.

The energy was not electric or magnetic, since he could have detected the presence of that kind of energy directly. It was not gravitational, since the gravitational potential of these machines —when measured as a function of their distance from the planet's center—had actually increased since he had first detected them. It was barely possible, of course, that some *primary* source beyond his detection-range might work on such a basis. But for the moment that hardly bothered him. It could be filed away for future reference.

There was almost certainly no direct mechanical link with a distant energy source. He felt sure that he would have *seen* any such, during his brief trip aloft.

Chemical energy, however, remained a distinct possibility. Normally—which usually meant, he reflected wryly, circumstances in which intelligence had not taken a hand—chemical reactions were too slow to provide useful energy, even though they were responsible for life. However, on a planet infested with such weirdly active carbon compounds, it would not do to be dogmatic on the matter.

It was known that reactions, in such circumstances, did go with enormous speed, though little actual quantitative work had been done on the matter of the energy involved. It was quite conceivable, in any case, that there might be some method of turning chemical directly into mechanical energy, without involving electricity as an intermediate stage.

Looked at from this viewpoint, several more possibilities as to the planet became evident. Its natives could survive, either by nature or intelligent adaptation, in an oxygen-rich atmosphere. Oxygen was one of the most virulently active elements in existence. Hence, it might not be too surprising to find such a people developing a chemical technology and bypassing the electricity a living creature should logically use—but wait. They had *not* bypassed electricity.

There were auxiliary machines, among the vehicles facing him, which did use it. Perhaps, these people had originally developed a normal technology, but, for some unaccountable reason, had never mastered space-flight! That was more than likely, if one assumed they did not merely tolerate oxygen, but *needed* it.

In that case, they would inevitably exhaust, in a relatively short time, the metal resources of a single planet.

They would be faced with the choice of developing machines that did not make demands on the metal supply, or of sinking to barbarism during the millions of years it would take new metal deposits to concentrate to usability.

This race might have succeeded in accomplishing the former—in which case, the exhaustion of the local ore veins could not be blamed on the poachers after all. The marauder might have planted the torpedoes in momentary pique, believing that a regular freighter had been there first and hoping to throw the production schedule of this planet out of step with that which had been recorded for it.

It was a very attractive idea, but the agent decided he should not go quite so far in pure speculation. There should be other possible sources of energy besides chemical activity, promising as such energy appeared to be. He could, for example, detect a pressure against his hull which seemed to be due to currents in the atmosphere. These must necessarily carry energy, though it seemed, at first estimate, that it could hardly be quantitatively adequate to run these machines.

There was nuclear energy. Obviously, these aliens did not use it directly, yet the possibility remained that it was their primary source and was stored in some non-self-destructive form within them. Strength was lent to this possibility by the presence of the ionization tube, which might well be used to locate radioactive materials. If, of course, the normal senses of the creatures were inadequate for the task. Atomic energy not under rigid control was always a rather frightening thing to contemplate, and he did not dwell on certain other unlikely possibilities concerning it.

He had already thought of solar energy, but had seen nothing to offset any of his earlier objections to this theory. On the whole, the chemical idea seemed the most worth following up.

He searched his memory for the little he knew about the high-speed chemical reactions of free-oxygen environments, and found a few helpful items. For one, they *did* involve solar energy—they employed it usually in breaking down water. The oxygen was freed to the surroundings, and the hydrogen combined with oxides of carbon to produce carbohydrates.

These, in turn, could react upon each other, with simple compounds and with some of the free oxygen, to produce incredibly complex substances whose detailed structure had never been worked out by any chemist of his people. This situation should, of course, result in a continual increase of free oxygen in the planet's atmosphere at the expense of the water.

Observation indicated that, actually, an equilibrium was usually attained in this respect. Whether the oxygen re-combined spontaneously with the hydrogen in the compounds, or whether still other high-speed reactions, of the same general type as the photosynthetic ones, did the trick, was still a matter of debate. Even the agent could understand, however, that the combination of oxygen with almost any of the complex carbon-hydrogen compounds would return the energy originally supplied by the sun.

If the compounds had any reasonable density, it should be possible to store quite a fuel supply in a very small space that way, using atmospheric oxygen to combine with it whenever desired. Even without precise figures, he felt sure that this would constitute an adequate energy-source for the machines he had been watching.

Was there anything he had overlooked? No—he was nothing if not thorough when he undertook a task of objective scientific analysis. A doer had his own pride to safeguard, and if he was not an intellectual in a strict sense, he did possess a first-rate mind.

How could this theory be checked experimentally? If it proved correct, there should be, somewhere on or within these machines, a store of hydrogen-carbon compounds. They should be absorbing atmospheric oxygen at a fairly high rate. And they should be exhausting water and, possibly, oxides of carbon.

He had no means for recognizing the hydrogen-carbon compounds, even if he found them, so there seemed little point in trying to take one of the mechanisms apart. No point even if its operator proved willing to allow it. However, there seemed to be a possible way of attacking the problem through the other facts. If an oxidizing reaction of the sort he had envisioned went on in a confined space, what would happen to the pressure? He pondered the problem.

Producing solid oxides would reduce pressure by removing oxygen. The formation of carbon dioxide would leave it unchanged, for there would be the same number of molecules after the reaction as before. Making water or carbon monoxide would give a pressure increase, since each molecule of oxygen would go into two molecules of the product.

All this, of course, assumed that water and the oxides of carbon were gases at this temperature. The method offered him two out of three chances of learning something—better, really, since it was likely that two, or all three, of the reactions occurred together. Only if CO_2 alone were produced, would there be a negative result. The catch seemed to be how one was to seal one of these devices in a gas-tight container, with a limited amount of atmosphere?

The container, of course, was available. His own ship had a good deal of waste space, left deliberately to allow for later modifications, if and when they were developed. He could open his hull for maintenance at virtually any point, and the openings were naturally designed to seal gas-tight, since his occupation was more than likely to lead him into corrosive atmospheres such as this.

He would have to be sure that he let the planet's air only into chambers where it could not reach either his own tissues or the ship's circuitry. No, wait. The test should take only minutes or hours, not years. Both his flesh and the silver wires could stand oxygen that long, and he could get rid of it later by opening the hull to the vacuum of space. That made matters easier—much easier.

But how could he detect the change in pressure, if it did occur? He did have manometers, of course. But they were vented to the outside of his hull. No one had foreseen a need for measuring internal pressure. He would have to do some more hard thinking.

What effects would pressure produce, besides merely mechanical ones? There would not be enough change, in the electrical properties of the exposed wires, for even the agent to detect. The change would probably not be fast enough to alter the temperature noticeably. And even if it did alter it, he would not be able to tell whether the change were due to gas laws, or simply the operation of the machine.

In the temperature range of this world, it was not really certain that all the products were gaseous, anyway. The mere fact that he had detected them in that form, during his approach, meant nothing. The infra-red spectrographic equipment he had used would have picked up trace quantities. It was unfortunate that its receivers were also aimed outward.

The agent could not, for the life of him, recall the vapor-pressure curves of any of the expected products—though, come to think of it, *something* was liquid here. The clouds he could see proved that, as did their precipitation on his half. He could not assume that it was one of the products he sought, however, and his best bet was still to maintain pressure change. If he could do it . . .

VIII

STUNNED, SHAKEN, the three humans stared at the star-traveler which had now so unbelievably and unexpectedly revealed itself in full. And the star-traveler stared back at them through its dull, opaque vision windows.

It was Candace Parsons who spoke first. "Why!" she exclaimed in a strained, oddly small voice. "Why—it looks like a gigantic bathysphere! Maybe . . ." she fell silent.

Hal Parsons, ignoring the rain that streamed down his face, said, "Maybe what, baby?"

"I don't know." Candace's voice remained off-pitch, tremulous. "I guess I was thinking that maybe—if he *is* from outer space —our atmosphere is like an ocean to him. Maybe he *is* a bathysphere."

"Why do you refer to that thing as *he?*" her husband asked sharply. "Whatever is inside probably has no more concept of sex as we know it than an amoeba."

"I don't know. I really don't know, Hal." Candace mopped the rainwater from her face with a khaki towel she had brought from the jeep. "I don't know, but *he* —just *seems* masculine somehow."

"If that's a bathysphere," put in Truck MacLaurie, with a forced attempt at levity, "I'd surer than hell hate to take a bath in it. How would I ever get out?"

"Truck!" said Candace, biting her underlip. "Don't you honestly know what a bathysphere is?"

"Isn't it a round bathtub?" Truck asked.

"For your information," Candace said, more to herself than to the young man who had blundered, "A bathysphere is a globular device designed by William Beebe for deep underwater observation. Professor Piccard later used an improved model to—"

Her husband, who kept his eyes riveted on the alien visitor, suddenly leapt at her and pushed her flat on her face against the hillside. As he did so he yelled at Truck, *"For God's sake, flatten out!"*

The alien was on the move. There could be no doubt about it this time. Candace, her face ashen, felt the near-earthquake vibration emanating from the advancing sphere and looked up, barely in time to see it zoom skyward, leaving boiling earth and mud in its wake.

The alien's rise was as rapid as the pursuit-foiling lifting processes attributed to flying saucers in the nation's press. He shot up a thousand feet—two thousand—and again they smelled the acrid aroma of metal heating up unbearably from friction with the atmosphere.

Feeling a sudden, shocking, incongruous disappointment, Candace cried, "Oh—he's getting away! *He's leaving Earth.*"

"No he isn't," said Truck, staring grimly up into the rain. "Get a load of *that!*"

That proved to be a sudden lateral maneuver on the part of the alien. It moved several hundred yards sideways and again was immobile. It was apparently as capable of remaining immobile in the atmosphere as it had been immediately following its self-burial in the rocky soil.

Candace could see the great round eyes, reduced to mere dots in the distance, trained steadily upon the three of them. She experienced paralyzing fear. It was obvious now that the alien failed to welcome close contact with humans, and was determined to resist investigation.

Secondarily—but no less frightening—was the thought that, being an alien, it could scarcely be expected to have humanitarian sympathies. It would probably be no more hesitant about wiping them out than most people were about destroying bothersome insects.

She glanced at her husband for reassurance, but saw in his fear-shadowed eyes a reflection of her own fears. She had learned, long ago—in high-school biology—that the legend of a snake's ability to paralyze a bird-victim with an hypnotic stare was utterly false. Yet Hal's trapped gaze failed to refute that ancient tale. His eyes remained fixed upon the strange object hovering almost motionless above them, half-veiled by a mist of its own creation.

Then, suddenly, Candace screamed. The alien was returning, swooping directly down toward them with the speed of a V-2. Before the echoes of her scream could dwindle and die away, it had landed—not upon them but in its former resting place. It perched there lightly, dominating the immediate landscape, its opaque twin lenses still fixed implacably upon them.

It was Harold, lifting himself slowly from the rain-soaked ground, who said, "Now I wonder just what in hell was the precise purpose of that maneuver."

Candace, close to hysteria from the backlash of terror and shock, replied, "You might just as well ask why such a creature does anything?"

"Funny thing," said Truck, brushing mud from the front of

his clothing. "I think it wanted a better look at us. Did you notice the way it kept those fish-eyes on us all the time it was dancing that rock-'n-roll over us?"

"I noticed," Harold Parsons replied tersely, his face still drained of its natural color. "What beats me is why it had to hop around like that."

Truck frowned at the looming bulk of the alien. Then he looked at his companions and rubbed the bristles on his chin. "Funny thing," he repeated. "I'm completely sure now it wasn't trying to scare us."

"Then just what do you think it *was* trying to do?" Candace asked.

Truck had latched on to something and, bulldog-like, he was not giving it up. "This probably won't make much sense to you eggheads," he told them, in his Southwestern drawl. "But the way that thing acted reminded me of an uncle of mine. His eyes aren't as good as they used to be, and he won't wear bi-focals. When he wants a good look at anything close-up, he has to pull his head back. Do you know what I mean?"

"If your uncle looks like that," said Candace, with a tremulous nod at the alien, "it's no wonder you're having trouble with your credits."

"Hold it, baby," said Hal, regarding MacLaurie with something like awe. "I think he's got something. Take a good look at those things our friend sees with—if seeing is what they're for. Its eyes are set at much too flat a curvature to enable it to see anything small and close up without some sort of focusing agent. I can detect no evidence of its having any. In that case . . ." He paused.

"You mean, I'm right?" Truck asked incredulously.

"I mean you could be," said Parsons. "Nice going, Truck."

He looked thoughtful for a moment, and then he added, "If it really *is* a space-traveling machine of some sort—and the evidence to date makes that highly probable—then its eyes would be designed for judging objects of immense size, immense distances away. It would need no focusing devices."

"All right, you two geniuses," said Candace, who had recovered a small measure of her equilibrium, "if it really is a space-traveler, why would it have to resort to such extremes just to get a good look at us? Surely it has all kinds of other senses—or instruments for measurement. If not, how could it have gotten here in the first place?"

Harold Parsons fished a limp cigarette from an equally limp pack in his breast pocket. He eyed it in disgust and quickly tossed it away. "Has it occurred to you, baby, that it may not be *that* simple?" he asked. "If its vision equipment is so faulty under Earth-conditions, it undoubtedly is faced with other problems."

He paused, wiped his forehead briefly dry, and added, "I'll stake my Ph.D. that we're just as big a problem to our friend as he is to us. We know that it is capable of radio communication by voice. But, so far, all that it has been able to communicate is the fact that it can indulge in parrot-like mockery of our speech."

"Hey!" said Truck, who had been listening attentively. "You mean it hasn't made sense out of what we were saying."

"What do *you* think?" said Parsons.

Candace said, "You know, this may be silly, but it makes me think of a movie I saw once—one in which an explorer on a strange island had to learn to get on with the natives by pointing out objects and then repeating over and over their speech equivalents. The natives had to do the same thing."

"You saw it once? I saw it six times," said Truck. "The guy kept pointing at trees and rocks, and describing them in English."

Hal Parsons threw the pack after his discarded cigarette. "Probably it was Robinson Crusoe!" he exploded. "But, once again, Truck, you and Candace could be on the nose. The only trouble is—I don't believe we managed to impart much information while our pal was zooming about." He paused, adding with a frown, "There's only one way to find out."

They plodded back to the jeep. Truck cranked the battery, while Parsons got the radio transmitter into operation. This time, he didn't have to speak first. The moment the receiver was working, he could hear his own voice coming through the earphones in a reiterated, "Who's that? *Hello!* . . . Who's that? *Hello?*"

Parsons acknowledged, with, "Hello out there. We were watching you just now."

Back it came. "Hello out there. We were watching you just now."

Infuriatingly, frustratingly, it went on—meaningless repetition following meaningless repetition. Finally, as before, Parsons had to give it up in disgust.

Candace produced some dry cigarettes from the expedition stores, and she and Hal smoked them silently, under the shelter of the jeep-top. Truck, who was in training, did not join them. It was a damp, disheartening breathing spell.

Finally, Candace said, "Well, remember Valley Forge. It's always darkest before the dawn."

"Frankly, I'd rather not think about Valley Forge right now," said Parsons unhappily. "If that thing isn't able to make sense out of us unless it sees us, and it *can't* see us—how in hell are we going to make sense out of it? I think we'd better get help from outside

—if we can. Okay, it's uphill for us again."

"Maybe not," said Truck. "Listen."

They heard the faint thrum of plane engines coming through the overcast, maintaining itself, growing louder. Parsons threw his cigarette away and said, "Come on, Truck. Let's get going."

It was the general again, anxious to know how they were making out. Parsons told him in terse syllables. Truck looked up from his battery-duty and said, "Getting anywhere, Sergeant?" And was rewarded by a shut-your-mouth gesture.

Parsons said, "I know it's tough. But you must be able to get through to us somehow. How about dropping a couple of philologists by 'chute?"

"We may have to," was the reply. "But only as a last resort. Blast this rain! But you're doing okay, Professor. Stay with it."

And that was that. It was a gloomy threesome that made its way slowly over the soggy hillside from jeep to alien. They walked slowly around the alien, and then stood in front of it, regarding a little more calmly now the disc-like, too-flat lenses that had gone opaque again.

"I wonder if it can see us at all from this distance," Parsons mused. Then, irrelevantly, "You wouldn't think, with all the resources of modern science *and* the Air Force,

they'd let a little rain stop them cold."

"It isn't a *little* rain," said Candace, who had been listening to her husband's colloquy with the general through one earphone. "It's a lot of rain—and it has raised hob all around here. The soil and rock formations aren't used to so much moisture. They just can't take it."

"Let's hope we can take it a while longer," said Parsons, putting an arm around her and squeezing.

"Don't, honey," she said. "I just can't take it right now."

"*Hey!*" called Truck, who had been eyeing the monster from a bit to one side. "Watch it! Something's happening!"

IX

As usual, the solution was ridiculously simple, once the traveler had thought of it. Most of the access-doors in the hull opened outward and all were operated electrically. He had perfect control over the current supplied to their operating motors. He knew that if he refrained from latching one or more of the doors, and simply held it shut with the motor, he could sense directly the amount of effort needed to keep it sealed against the internal pressure.

As far as he was concerned, it was a quantitative solution—if the pressure increased. If it decreased

—well, he would know it, from the extra effort needed to open the door. He was concentrating on immediate small details now—and very wisely.

With his machine, action could follow thought without delay. The moment he had his answer, a door swung open in the side of the great metal egg he was driving, and Earth's air poured in. Good as his seals were, the ship had not, of course, retained any significant amount of gas in the millennia it had been in space.

He did not bother to develop a plan for enticing one of the machines through the opening. He assumed, quite justly, that any intelligent mind must have a fair proportion of curiosity in its makeup. The fact that self-preservation might oppose this influence did not, as far as the agent knew or suspect, apply to the present situation. The risk of sacrificing even an expensive remote-controlled machine should be well worth taking in such circumstances. He simply waited for one of the devices to be driven into his ship.

Before this happened, however, there was a good deal of conversation among the machines present and, he presumed, the distant broadcaster—if, of course, it could be called conversation. The agent was still unable to reconcile this supposition with the absence of intelligent life in the present group.

At last, however, the expected event occurred. One of the machines swung about and moved toward the opening in the hull. Just outside, it halted, and the agent guessed at a brief burst of atmospheric pressure waves, though his manometers did not react fast enough to catch them. Then it entered.

It traveled on four struts instead of two. It became completely horizontal and advanced on the supporting struts. Evidently the upper ones, which the agent had seen, could be used for locomotion when desirable. Its entrance was slower than by its usual rate of motion, though the agent could not imagine why. The suggestion that slower motion made detail observation easier would never have occurred to a being whose perception and recording operations occupied fractions of a microsecond. Whatever the reason for the delay, it finally managed to get inside.

The agent wasted no time. Ready to observe anything and everything that resulted, he shut the access hatch.

Results, by his reaction-time standards, were slow—additional evidence that remote control was involved. The electromagnetic unit burst into activity the instant things finally began to happen. Some of the machines outside began to tap on the hull with dimly perceptible solid fragments, apparently pieces of silicate rock. The agent tried to find regularities in the blows that might be interpreted as communication code of some sort. He failed.

One of the devices, standing a little distance away, moved one of its attached fragments of metal until a hollow cylinder—which formed part of it—was in line with the hull. After a long moment the more distant end of the cylinder filled with gas, sufficiently ionized to be clearly perceptible to the alien.

The gas must have been under considerable pressure, for almost instantly it began to expand, driving before it a smaller fragment of metal which had plugged the tube. This fragment became progressively easier to perceive as its speed through the planet's magnetic field increased.

It emerged from the near end of the cylinder with sufficient momentum to continue in a nearly linear course, until it made contact with the hull. The agent watched with mounting excitement as it flattened, spread out and finally broke into many pieces. Incredible! He analyzed it, both electrically and mechanically, from the way it broke up. But he could make no sense of the operation.

After a time, the pounding ceased, and the two machines remaining outside drew together. No obvious activity came from them for some time.

Inside the hull, more interesting, possibly more understandable, events were taking place. The moment the door had closed, the machine trapped within had attempted to withdraw. Its action was a trifle faster than that of the ones still outside. The agent could not decide whether this meant that the escape reaction was automatic, or that a distant controller had turned his attention to the captive machine first.

It had pounded aggressively on the inside of the door in the same seemingly planless fashion as its fellows. Then it had slowed down, and began to move another of the strangely fashioned pieces of metal distributed about its frame. This abruptly became clearly perceptible, as an electric current began to flow through portions of its structure.

The source of the current was a seemingly endless supply of metallic ions—quite evidently chemical energy could be used for something. The current's function was less obvious, since it was led through a conductor whose greatest resistance was concentrated in a tight metal spiral.

This must in some way have been shielded from atmospheric oxygen, since, while it must have reached a fairly high temperature if the ion cloud around it meant anything, it nevertheless remained uncorroded. Heating the wire seemed all that the device accomplished—the agent refused to believe that the ion cloud was intense enough to help either in action or perception. The light and heat radiated were inconsiderable, but—*wait!* Perhaps that was it—perhaps *this* machine had eyes!

The agent examined the electrical device more closely, and discovered that part of its uncharged structure consisted of a roughly paraboloidal piece of metal, which must certainly have been able to focus light into a beam of sorts.

A few moments later, it became evident that it did just that. The agent's body was exposed in several places in this part of the ship, and, time after time, one part would be struck by radiance, while the rest were in more or less complete darkness. Furthermore, a few minutes' observation showed that when the machine moved at all it followed the direction in which the light beam happened to be pointing at the time.

Sometimes it did not move, though the beam kept roving around the chamber. The agent deduced from that one of two things. Either the device had several eyes, or the one it had was movable over virtually the entire sphere of possible directions. The thing was making an orderly survey of the interior of the space in which it was trapped. But it was carefully refraining from touching anything except the floor on which it stood.

That portions of this floor consisted of the agent's tissue made no difference to either party—as far as either knew. But the agent began to wonder how much of the exposed machinery of the ship would be comprehensible to the presumed distant observer.

Still more, he wondered how this presumed observer maintained contact with his machine. There was no energy whatever—in any form that the agent could detect—getting through his hull, either to or from the trapped machine. A minor exception to this might be the pressure waves generated by the stones striking his hull. But he had already failed to find in these blows any pattern at all, much less one which could be correlated with the actions of the machine inside.

Naturally, the thought that this might be an automatic device, similar to the mole robots, could hardly help occurring to the Conservationist. If this were the case, its present behavior was far more complicated than that of any such machine he had ever encountered. But hold on—he had already faced the implications inherent in that idea. So the technology of this world was more advanced, in some ways, than his own. There were still things the natives didn't know—things which would most certainly hurt them. Any concern he might have felt about himself was drowned in this larger solicitude.

He wondered whether he could so operate any of his own machinery to or through his prisoner, so as to convey a message of any sort. Certainly, if it used light as a vehicle of perception, it could detect motion on the part of the relays. For example—they were larger by quite a margin than the wave length of the radiation the hot wire was emitting in greatest strength.

There were several hundred thousand of them in the dozen square yards exposed to the direct-line vision of the captive, which should be enough to form some sort of pattern. Some sort of pattern, that is, if their owner could figure out how to operate them without making the ship misbehave.

He was still pondering this problem, along with the question of just what *would* be a meaningful pattern to the operators of the machine, when his attention was once more drawn to the outside.

The machines there seemed to have taken up a definite course of action. They had once more approached the hull, and were doing something to it which he could not at first quite understand. It quickly enough became evident, however. The brightness of the images he was receiving through the eyes, to which he had naturally been paying very little attention, began rapidly to decrease. Within a minute or so, the lenses ceased to transmit at all.

His tactile "sense" consisted in part of the ability to analyze the response of his hull to the vibrating impulses he applied to it. If such impulses were followed faithfully he could be sure that there was no mass in contact with the surface. On the other hand, if they were damped to any extent, he could form a fairly accurate idea of the amount and even some of the physical properties of such a mass.

In the present case, he discovered almost instantly that his eye lenses had been covered with a most peculiar substance. It not only adhered tenaciously to them, but seemed to absorb without noticeable reaction the same vibrations which had sent the soil dancing out of his way like summer chaff in a breeze. This did not particularly bother him, since the eyes were nearly useless for watching the machines anyway. But he kept trying to shake the material off, while he considered the implications of the move.

One was that the machines depended, far more heavily than he had suspected, on the sense of sight, and must suppose that he did likewise. Another was that they were about to take measures which they did not want observed by him. He did not worry seriously about anything they could do to his ship, but he began to listen very carefully for their footsteps all the same.

Another possibility was that they simply did not want him to fly away with the captive machine. To a race dependent upon sight, no doubt the idea of flying without it was unthinkable. He wondered, fleetingly, whether he should move a few hundred yards, just to see what effect the act had on them. Then the actions they were already performing caught his attention, and he shelved the notion. He became alarmed at what appeared to be an abrupt change of plan.

Two of the things were leaving the neighborhood, in a direction more or less toward the other electromagnetic radiator. Making allowances for the difficulty these machines apparently suffered in traveling over uneven terrain, the agent felt reasonably sure that this was their goal. The other two remained near him and settled down to relative motionlessness, as nearly as he could tell. He comforted himself with the thought that whatever plan they were attempting might demand some time to mature.

Perhaps the departing machines were going after additional equipment, though it appeared their goal might be attained more rapidly by sending other machines from the control point. However, it was quite possible that no others were available—such was likely enough to be the case on any of his own worlds, where only one individual in five hundred was machine-equipped, and over half of these were incapable of locomotion. Pride swelled in him at the thought, but he dismissed it as unworthy.

His soliloquy was interrupted by something that had not happened to him since his ship had first lifted from the world on which it had been built. The incident itself was minor, but its implications were not. The hull vibration, which he was still applying near all of his above-ground eyes, *stopped* near one of them.

He had not stopped it. The command for the carefully planned motion pattern was still flowing along his nerves. It should have been inducing the appropriate response in a fairly large group of relays. Something had gone wrong, and it produced a sudden crisis in his thinking.

The ship, of course, was equipped with a fantastic number of test-circuits, and he began to use them for all they were worth. It took him about three milliseconds to learn a significant fact. All the inoperative relays were close to, or actually within, the compartment where the captive machine was located. Closer checking showed that the trouble was mechanical—the tiny switches were being held in whatever position they had been in when the trouble struck.

Worse, the paralysis was spreading. It was spreading with a terrifying rapidity. The basic cause was not hard to guess, even with the details far from obvious. The agent instantly unsealed the door barring his captive from the outside world, and felt thankful that the controls involved still functioned.

The thing lost no time in getting out, and the pilot lost even less in getting the door securely sealed after it. For the time being, he completely ignored what went on outside, while he strove to remedy the weird disability. He was far from consoled by the thought, when it struck him, that he had proved what he wanted to know.

Something solid had blocked the relays—had, more accurately, *formed around* their microscopic moving parts. Whatever it was must have come in gas form for he would have felt the localized weight of a liquid, even inside. Most of the interior of his ship, as well as his own flesh, was still far colder than the planet on which he was lying.

Quite evidently one of the exhaust products of the captive machine, released as a gas, had frozen wherever it touched a cold surface. It might have been either water or one of the oxides of carbon. The agent neither knew nor cared. He proceeded to run as much current as possible through all his test-circuits, with the object of creating enough resistance-heat to evaporate the material.

The process took long enough to make him doubt seriously that his conclusion could be correct. But eventually the frozen relays began to come back into service. He could have speeded up the process, by going up a few miles and exposing his interior to the lowered pressure, and he knew enough physics to be aware of the fact.

It spoke strongly for the shock he had received that he never thought of this until evaporation was nearly complete. It was lucky for his peace of mind that he never realized what the liquid water formed in the process might have done to his circuits. Fortunately, formed as it had been, it contained virtually no dissolved electrolytes and caused no shorts.

He realized, suddenly, that he had permitted his attention to stray from the doings of the nearby machines for what might be an unwise length of time, and at once resumed his listening. Apparently, they were still doing nothing. No seismic impulses were originating in the area where he had last perceived them. That eased his mind a trifle, and he returned to the problem of the material covering his eyes.

This stuff seemed to be changing slightly in its properties. Its elasticity was increasing, for one thing, and the change seemed to be taking place more rapidly on the side from which the air currents were coming. The agent could think of no explanation for this. He tried differing vibration patterns on the stuff, manipulating them with the skill of an artist—but a long time passed before he had anything approaching success.

At last, however, a minute flake of the material cracked free and fell away—*and he could really see! He could* actually *make out what was going on!*

X

To UNDERSTAND what had gone on outside the alien to cause all this on a purely human plane, an observer of the whole would have had to go back to an earlier event entirely of Truck's doing.

As Truck spoke, something very definitely was happening to the visitor from outer space. Following the young athlete's pointing forefinger, the Parsons saw, with astonishment, that a section of the globular metal body was slowly, steadily opening—or was being opened.

It was circular, perhaps two feet in diameter, and its opening looked unexpectedly simple for a creature, or a machine, capable of interstellar flight. A section of the full, or outer body simply dropped open and outward—apparently on hinges.

"Like dropped underwear," Candace murmured, to be instantly quelled by a severely reproving look from her husband.

His expression remained firm. "I know what you're thinking," he told her. "It seems too simple. But consider this. Any alien using such a device on a strange world must be damned well capable of protecting itself."

"Maybe it's an airlock," suggested Truck.

"Maybe," said Hal Parsons, "but don't bet on it. It could be anything. We don't know enough about the nature of this—" He stopped, as Candace clutched his sleeve. "What is it, baby?" he demanded.

"Hal honey," she said, panic returning to envelope her like a torrent of water far colder than the rain. "Hal, honey, do you suppose it's coming out?"

"*It!*" Truck suggested. "Why not *them.* Why not some of those little green men that flyboy was talking about."

Parsons stared apprehensively at the opening in an effort to penetrate the darkness within. But he could see nothing—not even a shadow advancing toward them or hovering motionless in the gloom. He looked oddly at Truck and then began to lead his wife toward the jeep.

"Come on, Candace," he said.

"We'd better get the rifle from the trailer—just in case."

For an instant, Candace hesitated. She was a self-reliant, wholly modern girl, proud of her ability to handle herself as well as any man, in almost any situation. But her self-reliance crumbled when she looked again at the alien—huge, globular, impervious—with the ominous, gaping door part way up one of its flanks. This, obviously, was not a situation to be handled with reckless assurance.

She said, "Okay, honey," in a very meek voice.

Parsons said, "Better stick with us, Truck."

"I want to see what's going on," said MacLaurie, in his easy drawl. "Anyway, I don't figure our little pal here means any harm."

"Just how do you figure that?" Parsons asked sharply.

"If it was going to hurt us, it would have done so long before this," was Truck's sage reply.

"Don't be foolish, Truck," said Candace in an urgent tone. "It may have been merely softening us up before it opened that door."

Truck silenced her with, "I've a hunch you've been reading too many science-fiction stories lately, Candace."

"Hold tight then until we get back," Hal commanded. To his wife, in a lower voice he said, "I don't like leaving him here, either. But his mind's made up, and someone had better keep an eye on it."

"If that's *all* he does," murmured Candace.

"What's that?" her husband demanded. "What do you mean?"

"Nothing, honey," she said. But so great was her concern that she glanced several times over her shoulder while en route to the trailer. Fortunately for her peace of mind each time she looked the situation remained unchanged. Truck still stood there, his hands at his waist, his head cocked a little on one side as he regarded the menacing wide-open door.

"Better hurry, honey," she urged as they neared the jeep. "Something we can't cope with may happen any moment now!"

"So far, damn little has happened," grunted Parsons. "I'm beginning to wish it *would* do something menacing. This stalemate is getting on my nerves."

"I'm not so much worried about what *it* may do," said Candace. "At least, not right now. It's what Truck may do that's got me frightened."

Hal looked at her skeptically. But he speeded up his motions nevertheless. He got the canvas-covered Winchester out from under the trailer tarpaulin, stuffed a box of bullets into a pants' pocket and began hurrying back towards the hillside almost at a run.

They were two-thirds of the way towards their destination when

Candace, tagging and slipping a little at his heels, again gripped his arm convulsively and said, "Hal, he's going to do it. *He's going inside!*"

Parsons stopped dead in his tracks and yelled, *"Truck!* Stay where you are! Do you hear me? Don't go any nearer until we get there!"

As they watched, appalled, Truck MacLaurie looked over his shoulder at them. For a moment his grin flashed in the rain. Then moving with a deliberation that masked the speed he was employing—a trick his opponents on the football field had learned to rue, he moved directly toward the round, open door in the alien's flank, hoisted himself up to it, wriggled a moment or two and vanished inside.

A moment later, his deep voice rumbled at them through the rain. "I'm all right!" he shouted. "Don't worry!"

It was then that, without sound or warning, the open door in the alien's flank swung shut, sealing Truck inside.

Hal and Candace exchanged appalled glances and began to run toward the ship. Candace sprinted, stumbling and gasping, directly toward it. She would have hammered on the alien metal barrier with her fists had Hal not restrained her.

"Easy," he said in tones that suggested calmness maintained only by the greatest effort. "Easy,

baby. There's no sense of all of us walking into a trap until we see what can be done."

"But I can *hear* him!" she cried. And at that moment audible sounds of something banging on the inside of the alien trap could be heard.

"Hold it, honey," said Hal. He continued to restrain her until, finally, she gave up, her face white with horror beneath the mud that caked it. Then he picked up a couple of loose stones and fired them, hard, one after the other at the portion of the hull where the door had opened.

"I tried to tell you we shouldn't have left him," she burst out, looking wildly around for some stones to throw herself. "Honey, we're *responsible* for him. We should have made him come with us."

"It's a little late for that now, baby," said Parsons, breathing heavily as he let fly with another stone.

Inside the alien ship, Hal felt for a moment like a soft-bodied larval insect cruelly encased in a metallic cocoon. The impulse that had moved him to enter the door had been irresistible. It had occurred to him, even before the Parsons had given him his opportunity, that if an alien ship offering such an invitation took off unvisited he would regret it for the rest of his life.

More than anything else he was motivated by the thought of what a certain little red-headed coed back on the Montana Mines campus might have to say about it. Competition was heavy where that girl was concerned—and as far as Truck could see at the moment, running second would make life insupportable.

He had tried to remind himself of both the danger and idiocy of disobeying Parsons' warning. But —and this was true even with professors—Truck seldom troubled himself with the various levels of college teacherdom. Parsons, to Truck, was like most faculty members, tending to be overcautious about almost everything. A fine character, but *too damned careful.*

The door had been there, Truck was there—and the result had been as inevitable as Candace had foreseen. What Truck hadn't figured on was that his host would elect to slam the door on him so quickly.

Inside, it was dark—and it was cold. It was cold with a bone-chilling, impersonal quality that reminded the gladiator of the storage room in the Arizona meatpacking establishment where he'd held a summer job two seasons back. For one horrible moment he had the ghastly idea that he was undergoing some sort of deep-freeze process, following which he would be taken back to his chilly host's home planet, for thawing out and laboratory dissection.

A saving memory reminded

him that, minutes earlier, he had ribbed Candace unmercifully about her having read too many science-fiction magazines. Now, it appeared, the proverbial shoe was on the other foot with a vengeance—his own size thirteen. She might have read too many such stories, but he was living too many—one too many, to be exact.

But the vagrant whimsy restored what had become rather a shaky sanity—and a sane Truck MacLaurie, while not exactly a mental giant, was capable in an emergency of formidable thought and action. He realized that his surroundings, while unpleasantly cold, were not of a sufficiently low temperature to quick-freeze him. The process would last a long time. It might be unpleasant, but it offered further possibilities of escape.

He wondered what his surroundings looked like, and instantly remembered that he had stuffed a flashlight into his pants' pocket that very morning, in case he had to work the radio battery entirely under the jeep tarpaulin—to keep it from getting wet. In two seconds he had the flash out and turned on, and was surveying the strange cell in which he appeared to be imprisoned.

Earlier that year, one of his roommates, who was something of an electrical handyman, had taken apart an ailing television set in his fraternity house. Truck's brief glimpse of the seemingly endless and incomprehensible confusion of wires, in their pink insulation wrappers, had conjured up a vision of a beehive being invaded by an army of pink worms.

Now he derived somewhat the same impression—save that the worms appeared to be of white metal, either silver or platinum, and the confusion even greater. He bent over a sector of the complex wiring that looked vaguely familiar, then jumped as a thump sounded from outside the hull. It was quickly followed by another thump.

Good old Doc! he thought, and hammered back until his hand began to ache. He considered using the flashlight, then decided against it. The thumping stopped, and he wondered how Jonah had felt in the whale's belly, without even a flashlight.

Better keep moving, he told himself, as he felt the gooseflesh form on his forearms. *Better keep looking around. Better keep trying to make this whale sick enough to throw me up . . .*

Outside, Hal and Candace Parsons were engaged in grim activity, as Hal prepared to see what effect the rifle would have. "It won't do much good," he said somberly, slipping a bullet into the chamber. "I was figuring on using it more against what came out, if necessary, than against that solid beryllium egg, or whatever it is."

"Maybe you'd better not shoot," said Candace. "You might make it do something drastic. You might make it kill Truck, or take off with him."

"On the other hand," Hal said, trying to sight against one of the invisible hinges of the round trap-door in its flank, "I don't think I can hurt him much. But I just might annoy him into reopening that damned porthole."

He pulled the trigger, and they looked on, a bit desperately, as the steel-jacketed slug was shattered against the impervious hull. Somewhat to their relief, nothing happened. But there were no more thumps from inside the big globe.

"We've got to get help," said Hal quietly, returning the rifle to its canvas cover, before it could be damaged by the rain. "This situation has got out of hand. I don't care how many scientists break their skulls when they drop them through the cloud-layer. We can't stand by and leave Truck trapped in there."

"Of course we can't," said Candace. "I'm glad you feel so strongly about it. I was afraid he was getting on your nerves."

"Of course he was getting on my nerves," Hal Parsons said, somewhat testily. "But that doesn't mean I don't like the ham-handed . . ." He paused, finished casing the rifle, and added tersely. "Come on—let's get moving. Before we do, let's make sure our pal's eyes —if they *are* eyes—can't see what we're doing."

"How do you blindfold a giant baseball?" Candace asked.

"With whatever I can find at hand," said her husband. "You throw a pretty good stone. Let's see how you are at throwing mud."

He showed her what he had in mind, and there was plenty of mud in a hollow of the hillside that had been turned into a small muck-hole by the alien-induced deluge. It took them about five minutes before the "front" of the alien was well plastered, as well as its "eyes." When the job was done, they moved quickly back toward the jeep and the radio.

"You'll have to crank, baby," he told Candace. "I've been doing the talking to these characters, and there's no sense—"

"Of course." Candace cut him off. "Honey, I'm frightened. Do you suppose that thing has already—?"

She paused, and they both stopped walking. A brash, familiar voice had hailed them from a hundred yards to the rear. Unable to believe their ears, they exchanged a half-fearful glance, and then turned slowly toward the source of the sound. It was Truck, waving and coming toward them at a trot.

"I don't know exactly what happened!" was his answer to the question that burst from them both as he caught up with them. "All of a sudden—just a little while after your shot—he opened the door and I got out of there as fast as I could."

"What was it like?" Candace asked him. "It must have been horrible."

"I dunno," said Truck. "It was sort of interesting—but, brother, was it cold! *I damn near froze to death!*"

XI

THE REASON was obvious, of course. With an aperture of thirty centimeters and a focal length of about twenty-seven, the focus of the Conservationist's eye-lenses was highly critical; with the aperture about half a millimeter, as it had been left by the fragment of clay he had broken off, it became a minor matter.

He recognized the machines easily, near the edge of his new field of view, and began to work on the covering of a better-located eye. He did not succeed quite so well here, as the fragment he finally detached was larger, and the image correspondingly less clear, but it was still a good-enough job to enable him to follow the actions of the devices visually.

They were not traveling, as he has deduced already. Furthermore, a fourth machine, hitherto unnoticed, had joined them. All four had settled to the ground, so that their main frames took the weight normally carried by the traveling struts, which appeared merely to be propping the roughly cylindrical shapes in a more or less vertical attitude. The different ways in which this was accomplished, in different cases, did not surprise the agent. It would not have occurred to him to expect any two machines to be precisely alike, except perhaps in such standard subcomponents as relays. And it was, of course, fortunate that every new development happened in sequence, enabling him to analyze carefully as he went along.

The upper struts were moving rather aimlessly in general, but it did not take long for him to judge that their primary function was manipulation. The objects being handled at the moment were for the most part meaningless—apparently stones, bits of metal without obvious function, utterly unrecognizable objects which might be aggregates of the unfamiliar carbon compounds, though the agent knew no way to prove it. There were one or two exceptions. The device that had projected the slug of metal at his hull was easy to recognize, even though he had not perceived all of it at the time it was being used.

He tried to decide what parts of the machines functioned as their eyes, and was able to find them. It was not difficult, for no other portion was reasonably transparent. He discovered that all these vision organs were now turned toward him, but saw nothing surprising in the fact. The operators must have been familiar with the rest of the landscape, and did not expect anything of interest to show up on it.

Then the traveler noticed that all four of the machines were rising to their struts. As he watched, they began to move toward him.

At the same time, one of them extended a handling member toward a smaller fabrication, which almost immediately turned out to be another electromagnetic radiator. It was put to use at once, being swiftly raised to the upper part of the largest machine in the vicinity of the eyes, while a minor appendage of the handling limb which held it closed a switch.

This started the carrier frequency, after a delay which the agent was able to identify as due to the slow growth of the ion-clouds in portions of the apparatus —apparently they were produced by heating metal—and to the inherent lag of mechanical operations. The relays in the device were fantastically huge. They took whole milliseconds to operate and since they rather obviously had components consisting of multicrystalline pieces of metal, they must have had a sharply limited service life.

Evidently the natives had not gone far enough with metal technology even to get the most out of one world's supplies. This was a side-issue, however. A far more interesting development involved the modulation of the carrier. The agent found it possible actually to see the way this was being carried out.

An opening in the machine, not far below the eyes, rimmed with a remarkably flexible substance at whose nature he could only guess, began to open, shut and go through a series of changes of shape. He found it possible to correlate many of these contortions with the modulation of the electromagnetic signal. Apparently the opening was part of a device for generating pressure-wave patterns in the atmosphere.

The agent supposed that whatever plan the distant osbervers had been maturing must be moving into action, and he wondered what the machines were about to do. He was naturally a little surprised, since he had not expected any developments of this sort so soon.

Then he wondered still more, for the advance toward him which had been commenced halted, as suddenly as it had begun. Whatever had motivated them had either ceased—or the whole affair was part of an operation whose general nature was still obscure. It would be the better part of valor to assume the latter, he decided.

He watched all four of the machines with minute care. They were now balanced on their support struts. They were neither advancing nor retreating, and the upper members were moving in their usual random fashion. All eyes were still fixed on his ship.

Then he noticed that the pressure-wave assemblies of all four were functioning, although three did not possess any broadcaster whose signal could be modulated. He watched them in fascination. Sometimes—usually, in fact—only one would be generating waves. At others, two, three or all four would be doing so. Even the one with the broadcaster did not always have its main switch closed at such times. Something a little peculiar was definitely occurring.

It had already occurred to the agent that the atmospheric waves carried the control impulses for these machines. Why should the machines themselves be *emitting* them, however? Receivers should be enough for such machines. Then he recalled another of his passing thoughts, which might serve as an explanation. Perhaps there was only *one* operator for all of them. And after all, why not? It might be better to think of the whole group as a single machine.

In that case, the pressure waves, traveling among its components, might be coordination signals. They just might be. At any rate, some testing could be done along this line. Whatever limitations he and his ship might have on this world, he could at least set up pressure waves in its atmosphere. Perhaps he could take over actual control of one or more of these assemblies. He had had the idea earlier, in connection with radio-waves, and nothing much had come of it. But there seemed no reason not to try it again with sound. Nothing could surpass the experimental method when it was pursued with one strongly likely probability in mind.

A logical pattern to use would be the one that had been broadcast back to the distant observer a few moments before. It had been connected with a fairly simple, definite series of actions, and he had both heard and seen its production. He tried it, causing his hull to move in the complex pattern his memory had recorded a few seconds before. He tried it a second time.

"The thing's howling like a fire-siren!"

Just as when he had tried the same test with radio waves, there was no doubt that an effect had been produced, though it was not quite the effect the agent had hoped for. The handling appendages on all four of the things dropped whatever they were holding and snapped toward the upper part of their bodies. Once there, their flattened tips pressed firmly against the sides of the turrets on which their eyes were mounted.

For a monent, none of them produced any waves of its own. Then, the one with the broadcaster began to use it at great length. The agent wondered whether or not to attempt reproduction of the entire pattern it used this time, and decided against it. It was far more likely to be a report than involved in control. He decided to wait and see whether any other action ensued.

What did result might have been foreseen even by one as unfamiliar with mankind as the Conservationist. The machine with the broadcaster began producing more pressure waves, watching the ship as it did so. The agent realized, almost at once, that the controller was also experimenting. He regretted that he could not receive the waves directly, and wondered how he could make the other—or others—understand that their signals should be transmitted electromagnetically.

As a matter of fact, the agent could have detected the sound waves perfectly well, had it occurred to him to extend one of his seismic receptor-rods into the air. A sound wave carries little energy, and only a minute percentage of that little will pass into a solid from a gas. But an instrument capable of detecting the seismic disturbance set up by a walking man a dozen miles away is not going to be bothered by quantitative problems of that magnitude. However, this fact never dawned on the agent. Yet few would deny that he had done very well.

As it happened, no explanation was necessary for the hidden observer. He must have remembered, fairly quickly, that all the signals the agent had imitated had been radioed, and drawn the obvious conclusion. At any rate, the broadcaster was very shortly pressed into service again. A signal would be transmitted by radio, and the agent would promptly repeat it in sound waves.

Since the Conservationist had not the faintest idea of the significance of any of the signals this was not too helpful—but the native had a way around that. A machine advanced to the hull of the ship and scraped the clay from one of its eyes. The particular eye was the most conveniently located one, to the agent's annoyance. But fortunately it was not the only one through which he could see the things.

Then, an ordered attempt was begun, to provide him with data which would permit him to attach meanings to the various signal groups. Once he had grasped the significance of pointing, matters went merrily on for some time.

They pointed at rocks, mountains, the sun, each other—each had a different signal group, confirming the agent's earlier assumption that they were not identical devices. But there also seemed to be a general term which took them all in.

He was not quite sure whether this term stood for machines in general, or could be taken as implying that the devices present were part of a single assembly, as he had suspected earlier. While the lessons went on, two of them wandered about the valley seeking new objects to show him. One of these objects proved the spark for a very productive line of thought.

Its shape, when it was brought back and shown to him, was as indescribable as that of many other things he had been shown by them. Its color was bright green and the agent, perceiving a rather wider frequency band than was usable by human eyes, did not see it or think of it as a green object. He narrowed its classification down to a much finer degree.

He did not know the chemical nature of chlorophyll, but he had long since come to associate that particular reflection spectrum with photosynthesis. The thing did not seem to possess much rigidity. Its bulbous extensions sagged away from either side of the point where it was being supported. The handling extention that gripped it seemed to sink slightly into its substance.

He had never seen such a phenomenon elsewhere, and had no thought or symbol from the term *pulpy*. However, the concept itself rang a bell in his mind, for the machines facing him seemed fabricated from material of a rather similar texture. It was a peculiarity of their aspect that had been bothering him subconsciously ever since he had seen them moving. Now a nagging puzzlement — subconscious frustration was always unpleasant—was lifted from his mind.

The connection was not truly a logical one. Few new ideas have strictly logical connection with pre-existing knowledge. Imagination follows its own paths. Nevertheless, there *was* a connection, and, from the instant the thought occurred to him, the agent never doubted seriously that he was essentially correct. The natives of this planet did not merely use active carbon compounds as fuel for their machines. They constructed the machines themselves of the same sort of material!

Under the circumstances it was a reasonable thing to do—if one could succeed at it. The reactions of such chemicals were undoubtedly rapid enough to permit as speedy action as anyone could desire—at least as fast as careful thought could control. The agent's race had long since learned the dangers inherent in machines capable of responding to casual, fleeting thoughts and his ship's pickup-circuits were less sensitive, by far, than they might have been.

It was obvious why these devices were controlled from a distance, instead of being ridden by their operators, too. There must be some dangerous reactions, indeed, going on inside them. The agent decided it was just as well that his temporary prisoner had merely looked at the inside of his ship, without touching anything, and resolved to take no more such chances.

At any rate, there should be no more need for that sort of experiment. Language lessons were well under way. He had recorded a good collection of nouns, some verbs the machines had acted out, even an adjective or two. He was puzzled by the tremendous length of some of the signal groups, and suspected them of being descriptions, rather than individual basic words.

But even that theory had difficulties. The signal which, apparently, stood for the machines themselves, one which should logically have called for a rather long and detailed description, was actually one of the shortest—though even this took several hundred milliseconds to complete. The agent decided that there was no point in trying to deduce grammar rules. He could communicate with memorized symbols, and they would have to suffice.

Of course, the symbols that could be demonstrated on the spot were hardly adequate to explain the nature of Earth's danger. The Conservationist had long since decided just what he wished to say in that matter, and was waiting, impatiently, for enough words to let him say it.

It gradually became evident, however, that if he depended on chance alone to bring them into the lessons he was going to wait a long time. This meant little to him, personally. But the mole robots were not waiting for any instruction to be completed. They were burrowing on. The agent tried to think of some means for leading the lessons in the desired direction.

This took a good deal of imagination on his part, obvious as his final solution would seem to a human being. The idea of having to learn a language had been utterly strange to him, and he was still amazed at the ingenuity the natives showed, in devising a means for teaching one. It was some time before it occurred to him that *he* might very well perform some actions, just as *they* were doing. If he did *not* follow his own acts with signal groups of his own, these natives might not understand that he wanted theirs. The time had come for a more direct and audacious approach to the entire problem, and at the thought of what he was about to do his spirits soared.

He did it. He lifted the ship

a few feet into the air, settled back to show that he was not actually leaving, and then rose again. He waited, expectantly.

"Fly."

"Up."

"Rise."

"Go."

Each of the watching machines emitted a different signal, virtually simultaneously. Three of them came through very faintly, since the speakers were some distance from the radio. But he was able to correlate each with the lip-motions of its maker. He was not too much troubled by the fact that different signals were used. He was more interested in the evidence that a different individual was controlling each machine. This was a little confusing, in view of his earlier theories. But he stuck grimly to the problem at hand.

XII

HAL AND CANDACE Parsons, and Truck MacLaurie were sitting on a relatively mudless patch of earth, within comfortable watching distance of the alien. They had passed the saturation point in their general, rain-soaked misery, and the experience Truck had just been through had unnerved them all to the point where they desperately needed a rest.

Hal was putting Truck through something of a third degree. He was attempting to draw some specific information out of the athlete's unscholarly mind as to the precise nature of the alien's interior. It was proving to be rugged going, and his nerves were not in the best possible shape.

"Dammit!" he exploded, when Truck proved, for the twentieth time that he had no idea why he had been so suddenly allowed to leave. "The opportunity of the ages, and it has to be given a blockhead with an I.Q. of seventy-seven, who can't tell what it's all about!"

"Lay off him, honey," said Candace pointedly. "Truck's no blockhead. He's a blocking back, amongst other things. He just doesn't happen to be a scientist."

"Okay, if you say so." Hal ran unsteady fingers through his soaking-wet hair. "Sorry, Truck. It's just so infernally frustrating."

"Somebody's coming," said Truck with charitable forbearance, apparently unruffled by the catechism he had just been through. "Over there—look."

A muddy, heavily-encumbered figure was approaching them through the rain and mist. Catching sight of them, it waved.

Truck, rising, advanced along the hillside to meet it, while Hal and Candace rose slowly to their feet. On closer approach, it proved to be a soldier, mud-soaked and carrying a movie-camera slung over one shoulder, and what looked like a scintillometer over the other. Truck had quickly relieved the newcomer of a heavy walkie-talkie.

"Mr. and Mrs. Parsons?" the soldier said as he came up to them. "I'm General Wallace Eades. I've been talking to you upstairs long enough. I finally decided to make the drop myself."

"You don't know how glad we are to see you, General," said Candace, noting the two mud-dulled silver stars on the collar of his open shirt. "After three days with our friend over there"—she nodded toward the impassive, grey-metal globe—"we were beginning to wonder if we were humans ourselves."

General Eades, his blue eyes unusually bright and young and alert in his lined, leathery face looked at the monstrous bulk of the alien and stood for a moment in silent speculation. Then he said, "I was beginning to think it was all a pipe-dream. He's a big fellow, isn't he?"

For the next few minutes, he talked with Hal, letting the geologist brief him on recent events. Then, turning to Truck, "Quite an experience for you, young man. If we get out of this thing in any sort of shape, you'll be in Hollywood in ten days."

"Coach wouldn't like it," said the football player. "And I'm no Elvis Presley."

General Eades put his head back and laughed. Then he unslung the movie camera and said, "I gather you haven't made a pictorial record of your friend over there. I don't know about you, but I don't want to be laughed out of the service. I thought you said he only had two eyes. Isn't that a third? Did you put mud in that one, too?"

"I'll be damned!" said Hal. He and Candace regarded one another. They were bewildered, amazed and a little frightened. His lips tightening, Hal said, "He's full of surprises. Stick around and you'll find out."

"I intend to," said General Eades. "I've been on this thing, ever since the first radar flash came in—four days ago. Haven't had two hours consecutive sleep since. You've got no idea the fuss our friend has kicked up. The army's got ten thousand men trying to crack this valley, and diplomats and newspaper men are sleeping on billiard tables in Butte—if they're lucky enough to buy space on one."

As he spoke, he walked slowly around the monstrous globe, holding the camera to eye level, shooting it from all sides. Returning, he reappropriated the walkie-talkie from Truck, who had been dutifully standing guard over it.

"I checked the stuff in your jeep and trailer on the way here from my drop," General Eades said. "You must have got more than just arm-tired cranking that battery outfit of yours. I haven't seen one like it since World War Two."

"It was the best the department at the University could allow us," said Hal, a trifle on the defensive.

Tactfully, Candace put in, "We're awfully glad you got here, General. We were not only wet— we were lonely for a new face."

"Afraid mine's not exactly new," said Eades. Then, putting the walkie-talkie to work, he said resignedly. "Guess I'd better report, before they send a big drop in, and a few-score G.I.'s get killed. This valley's full of rocks and potholes, and visibility is nil."

"You're telling us, General!" said Truck.

The general's report, via radio, was lengthy but concise. He had yet to complete it when an audition from the alien, mimicking his own voice, caused interference that made intelligible communication impossible. He lowered the set, looked at the others, and nodded toward the grey-metal globe.

"Is that it?" he asked.

"That's it," said Hal.

Almost before the words were out of his mouth, a new sound— not through the radio, but carried clearly through the open air— smote all their ears. Smote was the word, as it rose in an ear-shattering crescendo that caused them to look at one another in alarm.

"The thing's howling like a fire-siren!" cried Candace, clapping her hands to her ears. The others followed suit.

It continued, for a couple of deafening minutes that all but reduced already quivering nerves to shreds. Then, as suddenly as it had started up, it ceased, and slowly they removed their hands.

Candace wondered if her ear-drums were permanently damaged. She saw Truck hammering the side of his head, like an inexperienced swimmer with water in his ear.

He said, "If that's his natural voice, I wonder how he sounds when he's *really* worked up."

Hal and the general exchanged a significant look. It was Eades who broke the welcome silence. "Maybe he's right," he said. "Is that the first time it's tried communicating—apart from radio mimicry?"

"That's right," Hal told him.

"Significant," said General Eades. "Damned significant. I wonder . . . That third eye bothers me. Do you suppose it bothers him?"

He walked up to the machine, disregarding Candace's gasp, "Be careful!"

Gently he scraped the mud from the lens. Nothing happened, but the sound did not return. He said, scowling at the porthole,

"The surface looks too flat for close vision."

"We had the same thought," Hal told him. "Still, it can see when it wants to."

General Eades walked around the sphere, studied the other two eyes, noted the places where the caked mud had flaked away. "Used to know an optometrist," he muttered. "Could be, the mud helps to give him closer focus by covering most of the lens."

". . . most of the lens," said the general, though his lips did not move. Eades started, looked at the others, and instantly pointed to one of his own eyes. He said, "Eye."

"Eye," said the voice from the alien. There was no question now in any of their minds. The alien had clearly discovered some means of direct vocal speech.

After several more tests, the general walked back to the others, his blue eyes alight with excitement. "That's it," he told them. "Our friend made that howl to let us know it had a *new* means of communication."

Hal motioned him to silence, and they waited, breathlessly. But the alien did not repeat the speech or any part of it. The geologist advanced, pointed to himself, and said, "Man."

"Man," said the alien.

"It understands," said Hal, his voice almost cracking. "Listen!" He accompanied the words by no pantomime, and the alien was silent.

"I'll be damned!" said the general.

"Eureka!" cried Candace, raising her arms toward the sky.

"Eureka!" said the voice from the globe.

"Careful, baby," Hal told her. "You just gave our friend a bum steer. Don't gesture unless you're outlining exactly what you mean."

From then on, in the excitement of attaining at least a rudimentary understanding with the thing from space, the little group on the hillside forgot the rain and their physical misery. Time was forgotten too, as they taught it new word meanings, and brought it examples of equipment and local flora in an effort to increase its vocabulary.

Candace found a bedraggled plant, wiped mud from it and said, "Green," pressing the stem as she held it up for the alien to see.

"Green," came the answer. "Plant—green."

"Green," Candace repeated. "Green through sun." She pointed skyward. "Green through photosynthesis."

"Plant green—through photosynthesis," came the expected reply. Then, "Plant, man green—both photosynthesis."

"Bless me!" cried General Eades. *"We're on the way!"*

Hal spoke up then. "Has it occurred to you, General, that our friend here may have some message to give us? If he has, it may take us a hell of a long time before we can give him the right words to give back to us."

Eades stroked his chin. "You were probably right in asking for philologists earlier," he said unhappily. "We're a bunch of babes in the woods at this game."

There was a long, disconsolate silence. Then Candace broke it, saying bravely, "In any case, we've got to keep going. Our friend may have an answer of his own."

"I'd give a lot for one good word-man I could count on getting down here alive," said the general. "I'll put in a call."

But, before he could get the radio in operation, the observant Truck said, *"Look!* Hey, don't tell me he's leaving us now!"

They stared in horror and utter dismay as the great, grey bulk of the alien rose vertically in the roil of mud already familiar to all but the general. Then they breathed sighs of relief. It hovered, only a few feet above the ground, then settled back, then rose again and remained stationary.

"He's trying to signal to us!" cried Candace, her voice shrill with excitement. "He wants us to give him a word for what he's doing."

"Fly!" shouted the general.

"Up!" said Candace.

"Rise!" called Hal Parsons.

"Go!" yelled Truck MacLaurie.

They spoke almost simultaneously, but the monster from space seemed confused. He made no answer at all.

XIII

THE AGENT dropped back to the ground and went through his actions again. This time only the individual with the radio spoke. The word it used was *Rise*. This was not the one it had used the other time. To make sure, the agent went through the act still again, and got the same word. Evidently, once their minds were made up, they intended to stick to their decisions. What *could* he think?

Then he tried burrowing into the ground, which seemed a useful action to be able to mention. The word given on the radio was *dig,* though two of the other machines apparently had different ideas once more.

It did not occur to him that these things might be detecting the by-products of his digging as well as his deliberate attempts to produce sound waves, or that his efforts to focus his third eye lens, a little while before, had actually been the cause of their sudden interest in his ship at that moment. He was much too pleased with himself at this point to entertain such extraneous ideas.

Having taken over the initiative in the matter of language lessons, he concentrated on the words he wanted, and, within a fairly short time, felt sure that he could get the basic facts of Earth's danger across to his listeners. After all, only four signal groups were involved in the concept. Satisfied that he had these correctly, he proceeded to use them together. In his progress now he felt the surge of a very personal kind of pride.

"Man dig—mountain rise."

For some unexplained reason the listening machines did not burst into frantic activity at the news. For a moment, he hoped that the controllers had turned to more suitable equipment to cope with the danger, leaving inactive that which they had been using. But he was quickly disabused of that bit of wishful thinking. The machine with the radio began to speak again.

"Man dig." It bent over and began to push the loose dirt aside with the flattened ends of its upper struts.

The agent realized, with some dismay, that its operator must suppose he was merely continuing the language lesson. He spoke again, more loudly, the two signal groups which the other seemed to be ignoring.

"Mountain rise."

All the machines looked at the hill across the valley, but nothing constructive seemed likely to come from that. If they waited for that one to rise noticeably, it would be too late to do anything about enlightening them as to the robots. He tried, frantically, to think of other words he had learned, or combinations which would serve his purpose. One seemed promising to him.

"Mountain break—Earth break —man break." The verb did not quite fit what was to happen, according to its earlier demonstration, but it did carry an implication of destruction, at least. His audience turned back to the ship, but gave no obvious sign of understanding.

He thought of another concept which might apply, but no word for it had yet appeared in the lessons. So, to illustrate it, he turned his ship's weapon on a patch of soil, a hundred yards from the bow. Twenty seconds' exposure to that needle of intolerable flame reduced the ground which it struck to smoking lava.

Even before he had finished, the word *fire* came from one of the watchers. The observer made no comment on the fact that the tube which threw slugs of metal had been leveled at his hull, during most of the performance. He simply made use of the new word.

"Man dig—Earth fire—mountain fire."

One of the machines produced its ionization tube and cautiously approached the patch of cooling slag. This had a slight amount of radioactivity from the beam, and its effect on the tube gave rise to much mutual signaling on the part of the machines. This culminated in a lengthy radio broadcast, not addressed to the agent. Then the language lessons were resumed, with the natives once more taking the initiative.

"Iron—copper—lead." Samples were shown individually.

"Metal." All the samples were shown together.

"Melt." This was demonstrated, when they finally made him understand that the weapon should be used again.

"Big—little." Pairs of stones, of cacti, coins and figures, scratched in the dirt, illustrated this contrast.

Numbers—no difficulty.

"Ship." This proved confusing, since the agent had supposed the word *man* covered any sort of machine.

Finally, slightly fuller sentences became possible.

"Fire-metal under ground," the men tried.

The agent repeated the statement, leaving them in doubt. More time passed, while *yes* and *no* were explained. Then the same phrase brought a response of "Yes."

"Men dig."

"Yes—men dig—mountain melt—mountain rise."

"Where?" This word took still more time, and was solved, at least, only by a pantomime involving all the men. *Here* and *there* were covered in the same act. However, knowing what the question meant did not make it much easier for the agent to answer it.

He had no maps of the planet, and would have recognized no man-made charts, with the possible exception of a globe, which is not standard equipment on a small field expedition.

After still more time, the men managed to get a unit of distance across to him, however, and he could use the ion beam for pointing. In this way, he did his best to indicate the locations of the moles.

"There! Eighty-one miles. Two miles down." And, in another direction. "There! Fifteen hundred-twelve miles. Eighteen miles down." He kept this up through the entire list of the forty-five moles he had detected and located.

The furious note-taking that accompanied his exposition did not mean anything to him, of course, though he deduced correctly the purpose of the magnetic compass one of the listening machines was using. He realized that giving positions to an accuracy of one mile was woefully inadequate for the problem of actually locating the moles.

But he could do the final close-guiding later, when the native machines approached their targets. He could come to their aid if they did not have detection equipment of their own which would work at that range. Just what possibilities in that direction might be inherent in organic engineering the agent could not guess. At any rate, the natives did not seem to feel greater precision was needed. They made no request for it.

In fact, they did not seem to want *anything* more. He had expected to spend a long time explaining the apparatus needed to intercept and derange the moles. But that aspect of the matter did not appear to bother the natives at all. Why, why? It *should* have bothered them.

In spite of appearances, the agent was not stupid. The problem of communicating with an intelligence not of his own race had never, as far as he knew, been faced by any of his people. He had tried to treat it as a scientific problem. It was hardly his fault that each phenomenon he encountered had infinitely more possible explanations than ordinary scientific observation, and he could hardly be expected to guess the reason why.

Even so, he realized it could not be considered a proven fact that the natives had read the proper meaning from his signaling. He actually doubted that they had, in about the way and to about the extent that some mid-nineteenth century human physicists doubted the laws of gravity and conservation of energy. He determined to continue checking as long as possible, to make sure that they *were* right.

The human beings, partly as a result of greater experience, partly for certain purely human reasons, also felt that a check was desirable. With their far better local background, they were the first to take action. To them, *fire metal,* when mentioned in conjunction with a positive test for radioactivity, implied only one kind of fire.

Man dig was not quite so certain. They apparently could not decide whether the alien being was giving information or advice —whether someone was already digging at the indicated points, or that they should go there themselves to dig. The majority inclined to the latter view.

To settle the question, one of them took the trench-shovel, which was part of their equipment, and arranged a skit that eventually made clear the difference between the continuative—*digging*—and the imperative *dig!*

While this was going on, another thought occurred to the agent. Since these things had used different words for the machines he was watching and the one he was riding, perhaps *man* was not quite the right term for the mole-robots he was trying to tell about. He wondered how he could generalize. By the end of the second run-through of the skit he had what he hoped was a solution.

"Man digging—ship digging," he said.

"Digging fire metal?"

"Man digging fire metal—ship digging fire metal."

"Where?"

He ran through the list of locations again, though somewhat at a loss for the reason it was needed, and was allowed to finish, because, though he did not know it, no one could think of a way to tell him to stop. He felt satisfied when he had finished—there could hardly be any doubt in the minds of his listeners now.

They were talking to each other again—the reason was now obvious enough. The operators must be in different locations, must be communicating with each other through their machines. He had little doubt of what they were saying, in a general way.

Which was too bad—in a general way.

"It's vague—infernally vague."

"I know—but what else can he mean?"

"Perhaps he's just telling about some of our own mines, asking what we get out of them or trying to tell us he wants some of it."

"But what can 'flame metal' mean but fissionables? And what mine of ours did he point out?"

"I don't know about all of his locations, but the first one he mentioned—the closest one—certainly fits."

"What?"

"Eighty-one miles, bearing thirty degrees magnetic. That's as close as you could ask to Anaconda, unless this map is haywire. There are certainly men digging there!"

"Not two miles down!"

"They will be, unless we find a substitute for copper."

"I still think this thing is telling us about beings of its own kind, who are lifting our fissionables. They could do it easily enough, if they dig the way this one does. I'm for at least calling up there, and finding out whether anyone has thought of drilling test cores under the mine level—and how deep they went. There's no point walking around here, looking for anything else. We've found our fireball, right here."

The agent was interested but not anxious when the machines turned back to him, and direct communication was brought once more into operation. He was beginning to feel less tense, and confident that everything was going to come out all right if he stuck with it.

"Eighty-one miles that way. Men digging. Go now."

They illustrated the last words, turning away from his ship and starting in the proper direction. The agent could not exactly relax, fitting as he did into the spaces designed for him in his ship, but he felt the appropriate emotion.

They were getting started on one of the necessary steps, at least. Presumably, the other and more distant ones would be tackled as soon as the news could be spread. These machines moved slowly, but their control impulses apparently did not.

It occurred to him that, since none of the devices had been left on hand to communicate with him, the natives might be expecting him to appear at the nearest digging site—the one they had mentioned. The more he thought of it, the more likely such an interpretation of their last message seemed. So, with the men barely started on their walk back to the waiting jeep, the Conservationist sent his ship whistling upward on a long slant toward the northeast.

XIV

THAT THE multiple answer had puzzled the star-traveler became evident when he dropped back to the ground and went through the entire process a second time. This time, General Eades took over, employing Hal Parsons' definition —"Rise."

Apparently, the switch in words from one member of the party to another troubled the alien, for he dropped gently. Then he rose and hung in the air once more, a few feet above the muddy soil. The general repeated, "Rise,"—and after a few seconds of motionless hovering the alien dropped back to the ground and did not go through the performance a fourth time.

"I wish our experimental boys would come up with an anti-grav like that," said the general, in a wistful aside. "It would sure give us the jump on you-know-who. Wonder where he gets all that power from."

"You and me both," murmured Truck MacLaurie. "What a bucking-machine he'd make for practice."

Candace giggled, and Hal looked at her, then despairingly at Truck. To give him solace, Candace said, "Just think, honey, the progress we've made in the last few hours. Only a little while back, we were nowhere."

"I wish I felt it were getting us anywhere," said the unhappy geologist. "When we started out looking for minerals, I never figured we'd come up against anything like this."

The general motioned them to silence, saying, "He's up to something new. Lordy! Just look at him *dig!*"

Candace said, "He's burrowing! That's all. What did you expect?"

Truck cried, "He's mining!"

"Will you shut up!" said Hal rudely, as General Eades scowled unhappily at the confusion-potential of another multiple answer. Another howling siren rose from the alien. But it was neither as enduring nor as ear-shattering as his earlier signal.

"Sorry, honey," Candace whispered when it was over.

It became quickly evident that the leadership in the language lesson had been reversed. Evidently having decided he had learned all he could from human demonstration, the visitor was demonstrating on his own, hoping the humans could supply the definitions he sought.

"I'd give my stars to know what he's trying to get through," said the general softly. "It must be important, if he's come all the way from God knows what star to give it to us."

There was another growl from the alien, which all four of them took as a request for silence. From then on, the reverse lesson went on apace. The only difficulty was that the words evidently sought by the visitor, made little sense to his watchers.

"Man dig—mountain rise," came the message.

They stared at one another, uncomprehending. Finally, with a shrug, the general bent over awkwardly, hampered as he was by the walkie-talkie, and began scraping a hole in the mud with his fingers.

"Man dig," Eades said, as he did so.

"Mountain rise."

There was insistence in the aliens words, which caused all four of his listeners to turn toward the ragged range-crests, which was barely visible now through the rain and mist on the far side of the valley. There was further confusion, when the great, grey globe gave voice to the strange words, "Mountain break—Earth break—man break."

Then, came sudden, unexpected, frightening demonstration in action. For the second time since its discovery an opening appeared in the dully gleaming, curved surface of the alien—an opening both smaller and more menacing than the one which had all but led Truck to his doom.

A snout appeared, swiveled past the watching group, and from it there emerged a darting, blinding ray of light—or heat. It struck the muddy hillside a hundred yards or so away and with frightful, eruptive violence a patch of the soggy soil itself began to bubble and turn, first red, then white-hot. A trickle of fluid, molten material ran slowly down the hillside, and a cloud of white steam rose high in the air. Once more, the sense of intolerable heat was present.

"My God!" exclaimed a white-lipped Hal Parsons. "He's set fire to the earth itself!" He picked up the rifle, which he was still carrying, unslung its cover and aimed it at the hull-opening, pushing

Candace behind him as he did so.

"Put that toy away, Parsons," said the general with grim insistence. "It won't do a damned bit of good. Do you understand? Put it away."

The intolerable ray of heat vanished, and the opening in the alien's hull disappeared as abruptly as it had opened. The visitor said, slowly, "Man dig—Earth fire—mountain fire."

"Let's have the scintillometer," the general said to Truck. "It's a lot better than that Geiger job you've been using."

"I can work it," said Hal. Taking the instrument, and adjusting it, he walked over to the rapidly cooling, but still semi-molten spot which the heat-ray had turned to lava. The count ran high and fast as he approached it. Turning back to the general, he said, "No doubt about it—she's plenty hot."

"Got to report this," said Eades tersely. "He's trying to get something through, all right—and *I don't like the looks of it*. Maybe some of those eggheads sitting around in Butte can give us a clue."

It was a lengthy broadcast, relayed through the radio of a helicopter hovering above the clouds. When it was over, the general signed off in disgust.

"How do you like that?" he said, to no one in particular. "Those broad-beamed boffins want us to carry on." He cursed, fluently, effectively, and then added, "Sorry, ma'am," to Candace without turning a grey hair.

She said, "Maybe we'd better try him on minerals alone."

So, the lesson continued, until some of the confusion about various stones and metals, upon the nature of machines, was partially cleared up. Then came the alarming statement, "Yes—men dig—mountain melt—mountain rise."

"Is he trying to tell us *men* are planting volcanoes under us?" Candace asked incredulously.

"He's trying to tell us someone or something is," her husband told her grimly. "Ask him where, General?"

This led to laborious exchanges, establishing direction and distance units, after which the alien began issuing his information, as to the location of the horrors to come if his warnings were ignored. While this was going on, Parsons took notes, doing his best to write legibly on limp paper. Finally Candace, who had once learned shorthand, took over the job.

General Eades turned toward her and said, "That seems to be all. Got them?"

"All forty-five, General," said Candace. "Want me to read them back to you?"

"Not yet," said Eades. "I want to know what he means by *men* digging volcanoes."

The results were not satisfactory, and so the alien went through the entire list again. Then, in desperation, the general got into touch with the higher-ups once more. He talked long and determinedly and with authority. He concluded with, ". . . There's absolutely no point in walking around here, looking for anything else. We've found our fireball, right here."

He paused, looked at the impassive facade of the alien inquiringly. "More?" he asked. "Anything more?"

The voice, so oddly human, so utterly like his own in tone and inflection, replied, "Eighty-one miles that way. Men digging. Go now."

"Okay," said the general. "That's the message." And, to the alien, "Eighty-one miles that way. Men digging. Go now." He motioned to the others to follow, and led the way through the rain toward the jeep.

"You're not going to leave him?" Candace asked, incredulously.

"It may take all of us to get out of this damned valley," Eades told her. "If what he reported is true—no matter how garbled— our work is at Anaconda. That's where the nearest trouble is, according to him. We'll have weasels in here by tomorrow, to do a proper survey job. Complete with scients . . ." Then, with a look of apology, "Sorry, folks, I mean specialists. You've done great."

"That's okay, General," said Truck, in his easy-going drawl.

The others laughed.

Candace said, "This probably sounds screwy, but I'm going to miss our globular friend. He was—"

"Not *he—it*," said Parsons. "Why must you give it *sex?*"

"Forget about sex," the general told them, masking a smile. "We're going to have one sweet job getting out of here."

Candace looked back, through the mist and rain and darkness of approaching twilight, and suddenly uttered a cry of alarm. "Look!" she said, grabbing the nearest arm, which happened to belong to the general. "He's taking off!"

They watched, all with mixed emotions, as the alien rose vertically from its hillside bed, and hovered a moment at mountaintop level. Then it suddenly veered, moved swiftly toward the north and disappeared.

"Well," said Truck. "Goodby."

And that seemed to sum it up. Before they had the jeep halfway up the pass the rain had stopped, and there was a break of afterglow gold in the western sky.

XV

THE MOMENT HE rose above the valley, the Conservationist picked up the radar beams again —the beams that had startled him when he first approached the strange planet. As had happened on the earlier occasion, a few milliseconds served to bring many more of them to bear upon him.

He was quite evidently being watched on this journey. But he no longer expected these beams to carry intelligent speech. More or less casually, he noted their points of origin. He wondered, for brief moments, whether it might not be worth while to investigate them later, but felt fairly certain that it wouldn't. He turned his full attention on his goal.

The crusts of clay had fallen from his eyes as he flew, and he was once again limited to long-distance vision. He could make out the vast, terraced pits of the great copper mine as he approached, but could not distinguish the precise nature of the moving objects within. He did not consider sight a particularly useful or convenient sense anyway, so he settled to the ground, half a mile from the pit's edge, bored in as he had before, and began probing with seismic detectors and electrical senses.

He had, of course, already known of the presence of the hole. A fair amount of seismic activity had reached his original landing-spot from this place, enabling him to deduce its shape fairly accurately. Now, however, he realized —and for the first time—the amount of actual work going on. There were many machines of the sort he had already seen, which was hardly surprising. But there were many others as well, and the fact that most of them were metallic in construction startled him considerably.

There was a good deal of electrical activity, and at first he had hopes of finding an actual native. But these hopes quickly faded when he discovered there was nothing at all suggestive of thought-patterns. Some of the machines were magnetically driven. Others used regular electrical impulses for, apparently, starting the chemical reactions which furnished their main supply of energy.

The really surprising fact was the depth of the pit. If this work had begun since the receipt of his information, the wretched, guilty robots would be caught without difficulty. It took some time, by his perception standards, for a truer picture of the situation to be forced on his mind.

The pit had *not* been started recently. The progress of the diggers was fantastically slow. Clumsy metal scoops raised a few tons of material at a time and deposited it in mobile containers that bore it swiftly away. Fragments of the pit-wall were periodically knocked loose by expanding clouds of ionized gas, apparently formed chemically. The shocks initiated by these clouds were apparently the origin of most of the temblors he had felt from this source, while he was still eighty miles away.

His electrical analysis finally gave him the startling, incredible facts. This was a copper mine—extracting ore far poorer in quality than any his own people could afford to process. This race was certainly confined, for some reason, to its home planet, and had been driven to picking leaner and ever leaner ores to maintain its civilization.

The development of organic machines had given them a reprieve from barbarism and final extinction, but surely could not save them forever. *Why in the galaxy,* did they not use the organic robots for digging directly, as he had seen them do, during the language lessons? One would think that metal would be far too precious to such planet-bound people, for them to waste even iron on bulky, clumsy devices such as those at work here!

Even granting that the machines he had originally seen, and which seemed the most numerous, were not ideally designed for excavation work, surely, surely, better ones could be made. A race that could do what this race had done with carbon compounds could have no lack of ingenuity—or, more properly, of creative genius.

Very slowly, he realized why they had not—and why his mission was futile. He realized why these people would be doomed, even if the moles had never been planted. He noticed something relevant, during the conversation, but had missed its full staggering implication. The organic compounds were *soft.* They bent and sagged and yielded to every sort of external mechanical influence—it was a wonder, thinking about it, that the machines he had seen held their shapes so well. No doubt, there was a framework of some sort, perhaps partly metallic even though he had not perceived it.

But such things could never force their way through rock. The only way they *could* dig was with the aid of metallic auxiliaries—simple ones, such as those used to illustrate the verb to him, or more capacious and complex ones like those in use here.

This race was doomed, had been doomed long before the poachers ever approached their planet. They needed metal, as any civilization did. They were bound to their world, but kept from moving about even upon it, for not one in a thousand of these people could conceivably travel by machine, as the agent's race did. The organic engines could not possibly be used as vehicles. They could not be so used because their very essential nature of chemical violence made them *untouchable.*

These people were trapped in

a vicious circle, using their metal to dig more metal, sparing what little they could for electrical machinery and other equipment essential to a civilization, always having less and less to spare, always using more and more to get it. The idea that they could survive, until the planet's natural processes renewed the supply, was ridiculous.

It was, in short, precisely the same tragic circle that the agent's own race was precariously avoiding, millennium after millennium, by its complex schedule of freighters that distributed the metal from each planet in turn among thousands of others, then either waited for nature to renew the supply, or "tickled up" uninhabitable worlds as the poachers had done to this one.

Metal kept the machines operating. The machines kept food flowing to that vast majority of individuals who could not travel in search of it. A single break in the transport schedule could starve a dozen worlds. It was a fragile system, at best, and no member of the race liked to think about—much less actually face—examples of its failure.

The agent's mounting discomfort as he considered the matter of Earth was natural and inevitable. This race was what his own might have been, hundreds of millions of years before, had means of space-travel not been developed. They would probably be extinct before the poachers' torpedoes began to take effect, which was, no doubt, a mercy.

The agent could not help them. Even if the communication problem were cracked, they could not be brought into the transport network of civilization for untold millennia. No, they were truly lost—a race under sentence of extinction. The reorganization necessary was *frightening* in its complexity, even to him. Teaching them to build and use the equipment of his ship would be utterly useless, since it was entirely metallic, and they would be even worse off than with their organic devices.

They were already, probably by chemical means, stripping ores more efficiently than his own people, so he could hardly help them there. No, it was a virtual certainty that, when the planet's crust began to heave as giant bathyliths built up beneath it, when rivers of lava poured from vents scattered over the planet, no one would be there to face it.

This was a relief, in a way. The agent could picture, all too vividly, the plight of seeing a close friend engulfed only a few miles away, and having to spend hours or years of uncertainty, wondering when his own area would be taken —*and then knowing.*

That was the worst. There was plenty of warning, as far as awareness was concerned. Anywhere from minutes to years and millennia, if one was a really good computer. You knew, and if you had a mobile machine, you could move out of the way. Even these organic machines traveled fast enough for that. But *only* machines would let a being get out of the way—and there would be no machines here by then.

He wished with every atom of his being that he had never detected the poachers, had never seen this unfortunate planet or heard of its race. No good had come of it—or very little, anyway. There would, admittedly, be metal here before long, brought up with the magma flows, borne by subcrustal convection-currents in the stress-fluid that formed most of the worlds bulk.

The poachers would be coming back for it, and he could at least deprive them of that. He would beam a report in toward the heart of the galaxy, making sure it did not radiate in the direction they had taken. Then there would be freighters to forestall them.

It was ironic, in a way. If any of this race should have survived the disturbance that would bring back the metal, that disturbance would be the salvation both of their species and their civilization. Most probably, however, the only witnesses would be a few half-starved, dull-minded barbarians, who would wonder, dimly, what was happening for a little while before temblors shattered their bodies forever.

There was nothing to keep him here, and the place was distasteful. More of the organic robots were approaching his position, but he did not want to talk any more. He wanted to forget this planet, to blot the memory of it forever from his mind.

With abrupt determination, he sent the dirt boiling away from his hull in a rising cloud of dust, pointed his vessel's blunt nose into the zenith and applied the drive. He held back just enough to keep his hull temperature within safe limits, while he was still in the atmosphere.

Then, with detectors fanning out ahead, he swung back to the line of his patrol orbit, and began accelerating away from the Solar system. Ignorant of events behind him, he never sensed the flight of swept-winged metal machines that hurtled close below while he was still in the air, split seconds after he had left the ground.

He did not notice the extra radar beam that fastened itself on his hull, while the machine projecting it flung itself through the sky, computing an interception course. This was too bad, for the relays in that machine would have made him feel quite at home, and its propulsion mechanism would have given him more food for thought.

He might have sensed its detonation, for his pursuer had a nuclear warhead. But its built-in brain realized, as quickly as the agent himself could have, that no interception was possible within its performance limits. It gave up, shutting off its fuel and curving back toward its launching station. Even the aluminum alloys in its hull would have interested the agent greatly—but he was trying to think of anything except Earth, its inhabitants and their *appalling* technology.

His patrol orbit would carry him back to this vicinity in half a million years or so. The freighters would have been there by that time.

He wondered if he could bring himself to look at the dead world.

IT WAS THE general who explained it to the Parsons, at the University a few weeks later. He said, "He must have been in the devil's own hurry. All he did was get his warning through, take a quick look at Anaconda, and zoom off. Ground-to-Air sent up a nuclear rocket to intercept him, but he got clear of it just in time, thank God! Plenty of heads rolled after that foul-up, I can assure you. Trigger-happy idiots they were!"

Candace, looking exceptionally attractive in a new, soft-blue linen dress which almost miraculously complemented both her figure and her coloring, said, "I'm glad, too. It must have had something to do with his intuitive alertness, from what I've been able to gather. Perhaps, he thought this world was going to blow up at any minute."

"Hah!" said General Eades. "We've already located nine of those damned underground borers he told us about. At the rate *they're* moving, our fiftieth-generation descendants will be out in space themselves before anything catastrophic happens. We'll have the whole bunch spotted and disarmed by that time."

He paused, chuckled again and added, "The weird part of it is that twenty-seven of the damned monsters are doing their stuff under Iron Curtain soil."

Hal Parsons spoke thoughtfully. "I've been reading some of the pull-together reactions in the headlines, General. Won't all this put you out of a job?"

"Not for a while," said Eades. "Actually, I hope so. No responsible soldier wants war—ever. Makes our uniforms too dusty."

"I still wish I knew how he produced that rain," said Candace. "I've added meteorology to my other duties, hoping to get to the bottom of it."

"Probably, he was just taking a bath," said Eades. He puffed on his cigar meditatively and added, "It's good to know you got a full professorship out of it, Parsons—and that you're on your way to one yourself, Mrs. Parsons." He fingered the new, bright extra star on his own collar, then asked, "What happened to the big, good-looking kid you had with you? I thought for sure he'd be in Hollywood by now."

"Oh—poor Truck," replied Candace. "He was all set to go. But he wanted to play in the homecoming game first. He broke his nose, and right now the movie brass isn't interested. But he doesn't seem to mind. He's making out fine with one of those cute little red-headed co-eds on the campus."

"I'm glad to hear it," said the general. He paused, frowning. "You know, it's funny—but ever since that damned metal monster flew out of our lives, I feel as if I'd lost a friend."

"I feel the same way," said Hal.

"I guess we all do," said Candace. She was much too wise, being a woman, to add, "I told you so."

Headline Novel — Next Issue

•

OPERATION: SQUARE PEG

by IRVING W. LANDE and FRANK BELKNAP LONG

On the tumultuous frontiers of tomorrow men will be judged as never before by their strongest psychological drives. And that's why there's thunderclap entertainment when war and science chart a new world.

Complete IN THE APRIL ISSUE

THE LAST WORD

It was the dreaded moment of Lucifer's victory over Man. The world lay ruined by war—but one human pair still lived.

by DAMON KNIGHT

THE FIRST WORD, I like to think, was "Ouch!" Some caveman, trying to knock a stone into a closer fit with another stone, hit his thumb—and there you were—language!

I have an affection for these useless and unverifiable facts. Take the first dog—he, I feel sure, was an unusually clever but cowardly wolf who managed to terrorize early Man into throwing him a scrap. Early man was a terrible coward, himself. Man and wolf discovered that they could hunt together, in their cowardly fashion, and there you were again—domesticated animals!

I admit that I was lax during the first few thousand years. By the time I realized that Man needed closer supervision, many of the crucial events had already taken place. I was then a young—well, let us say a young fallen angel. Had I been older and more experienced, history would have turned out very differently.

There was that time when I happened across a young Egyptian and his wife, sitting on a stone near the bank of the Nile. They looked glum—the water was rising. A hungry jackal skulked not far away, and it crossed my mind that, if I distracted the young people's attention for a few minutes, the jackal might surprise them.

"High enough for you?" I asked agreeably, pointing to the water.

They looked at me rather sharply. I had put on the appearance of a human being, as nearly as possible, but the illusion was no good without a large cloak, which was too hot for the time of year.

The man said, "If it never got any higher, it would suit me."

"Why, I'm surprised to hear you say that!" I replied. "If the river didn't rise, your fields wouldn't be so fertile—isn't that right?"

"True," said the man, "but also, if it didn't rise, my fields would still be my fields." He showed me where the water was carrying away his fences. "Every year, we argue over the boundaries, after the flood, and, this year, my neighbor has a cousin living with him. The cousin is a big, unnecessarily muscular man." Broodingly, he began to draw lines in the dirt with a long stick.

These lines made me a little nervous. The Sumerians, up north, had recently discovered the art of writing, and I was still suffering from the shock.

"Well, life is struggle," I told the man soothingly. "Eat or be eaten. Let the strong win, and the weak go to the wall."

The man did not seem to be listening. "If there was some way," he said, staring at his marks, "that we could keep tally of the fences, and put them back exactly the way they were before—"

"Nonsense," I interrupted. "You're a wicked boy to suggest such a thing. What would your old dad say? Whatever was good enough for him . . ."

All this time, the woman had not spoken. Now she took the long stick out of the man's hand and examined it curiously.

"But why not?" she said, pointing to the lines in the dirt. The man had drawn an outline roughly like that of his fields, with the stone marking one corner.

It was at that moment that the jackal charged. He was gaunt and desperate, and his jaws were full of sharp yellow teeth.

With the stick she was holding, the woman hit him over the snout. The jackal ran away, howling piteously.

"Tut!" I said, taken aback. "Life is struggle—"

The woman said a rude word, and the man came at me with a certain light in his eye, so I went away. When I came back after the next flood, they were measuring off the fields with ropes and poles.

Cowardice again—that man did not want to argue about the boundaries with his neighbor's muscular cousin. Another lucky accident, and there you were—geometry!

If only I had had the foresight to send a cave bear after the first man who showed that original, lamentable spark of curiosity . . . Well, it was no use wishing. Not even *I* could turn the clock back.

Oh, I gained a few points as time went on. Instead of trying to suppress the inventive habit, I learned to direct it along useful lines. I was instrumental in teaching the Chinese how to make gunpowder—seventy-five parts saltpeter, thirteen parts brimstone, twelve parts charcoal, if you're interested. But the grinding and mixing are terribly difficult—they never would have worked it out by themselves.

When they used it only for fireworks, I didn't give up—I introduced it again in Europe. Patience was my long suit—I never took offense. When Luther threw an inkwell at me. I was not discouraged. I persevered.

I did not worry about my occasional setbacks—strangely, it was my successes that threatened to overthrow me. After each of my wars, there was an impulse that drew men closer together. Little groups fought each other until they formed bigger groups—then the big groups fought each other until there was only one left.

I had played this game out over and over, with the Egyptians, the Persians, the Greeks, and, in the end, I had destroyed every one. But I knew the danger. When the last two groups spanned the world between them, the last war might end in universal peace, because there would be no one left to fight.

My final war would have to be fought with weapons so devastating, so unprecedently awful, that man would never recover from it.

It was so fought.

On the fifth day, riding the gale, I could look down on a planet stripped of its forests, its fields, even its topsoil. There was nothing left, but the bare, riven rock, cratered like the Moon. The sky shed a sickly purple light, full of lightnings that flickered like serpents' tongues. Well, I had paid a heavy price, but Man was gone.

Not quite, as it developed—there were two left, a man and a woman. I found them, alive and healthy for the time being, on a crag that overhung the radioactive ocean. They were inside a transparent dome, or field of force, that kept out the contaminated air.

You see how nearly I had come to final defeat? If they had managed to distribute that machine widely before my war started . . . But this was the only one they had made. And there they were, inside it, like two white mice in a cage.

They recognized me immediately. The woman was young and comely, as they go. She was called Ava Something-or-other—I didn't catch her husband's name. Gardiner, or something like that.

"This is quite an ingenious device," I told them courteously. In actuality, it was an ugly thing, all wires and tubes and so on, packed layers deep under the floor, with a big semicircular control board and a lot of flashing lights. "It's a pity I didn't know about it earlier—we might have put it to some use."

"Not this one," said the man

grimly. "This is a machine for peace. Just incidentally, it generates a field that will keep out an atomic explosion."

"Why do you say, 'just incidentally?' " I asked him.

"It's only the way he talks," the woman said quickly. "If you had held off another six months, we might have beaten you. But now, I suppose, you think you've won."

"Oh, indeed!" I said. "That is, I will have, before long. Meanwhile, we might as well make ourselves comfortable."

They were standing in tense, defensive attitudes, in front of the control board, and took no notice of my suggestion.

"Why do you say I 'think' I've won?" I asked.

"It's just the way I talk," she replied. "Well, at least, we gave you a long fight of it."

The man put in, "And now you're brave enough to show yourself." He had a truculent jaw. There had been a good many like him in the assault planes, on the first day of the war.

"Oh," I said, "I've been here all the time."

"From the very beginning?" the woman asked.

I bowed to her. "Almost," I said, to be strictly fair.

There was a little silence, one of those uncomfortable pauses that interrupt the pleasantest of talks. A tendril of glowing spray sprang up just outside. After a moment, the floor settled slightly under us.

The man and woman looked anxiously at their control board. The colored lights were flashing.

"Is that the accumulators?" I heard the woman ask in a strained, low voice.

"No," the man answered, "they're all right—still charging. Give them another minute."

The woman turned to me. I was glad of this, because there was something about their talk together that disturbed me. She said, "Why couldn't you let things alone? Heaven knows we weren't perfect, but we weren't *that* bad. You didn't have to make us do that to each other."

I smiled. The man said slowly, "Peace would have poisoned him. He would have shriveled up like a dried apple."

It was the truth, or near enough, and I did not contradict him. The floor lurched again.

"You're waiting to watch us suffer," the woman said. "Aren't you?" I smiled. She added, "But that may take a long time. Even if we fall into the ocean, this globe will keep us alive. We might be in here for months before our food gives out."

"I can wait," I said pleasantly.

She turned to her husband. "Then we *must* be the last," she said. "Don't you see? If we weren't, would he be here?"

"That's right," said the man, with a note in his voice that I did not like. He bent over the control board. "There's nothing more to keep us here. Ava, will you . . . ?" He stepped back, indicating a large red-handled switch.

The woman stepped over and put her hand on it.

"One moment," I said uneasily. "What are you *doing?* What *is* that thing?"

She smiled at me. "This isn't just a machine to generate a force field," she said.

"No?" I asked. "What else?"

"It's a Time Machine," the man said.

"We're going back," the woman whispered, "to the beginning."

Back, to the beginning, to start all over.

Without me, they were going.

The woman said, "You've won Armageddon, but you've lost Earth."

I knew the answer to that, of course, but she was a woman and had the last word and I let her have it.

I gestured toward the purple darkness outside. "Lost Earth? What do you call this?"

She poised, her hand on the switch. "Hell," she said.

I have remembered her voice as she said it, through ten thousand lonely years.

"The Magazine that is a Book!"

BRINGS YOU AN UNUSUAL SCIENCE FICTION COMPLETE NOVEL IN THE NEXT ISSUE

●

By IRVING W. LANDE and FRANK BELKNAP LONG

THE NEXT TENANTS

We can expect to hear odd confessions over a bar. However, mercifully seldom are they as strange or quite as terrifying as this one.

by ARTHUR C. CLARKE

THE NUMBER of mad scientists who wish to conquer the world," said Harry Purvis, looking thoughtfully at his beer, "has been grossly exaggerated. In fact, I can remember encountering only a single one."

"Then there couldn't have been many others," commented Bill Temple, a little acidly. "It's not the sort of thing one would be likely to forget."

"I suppose not," replied Harry, with that air of irrefragible innocence which is so disconcerting to his critics. "And, as a matter of fact, this scientist wasn't really mad. There was no doubt, though, that he was out to conquer the world. Or, if you want to be really precise—to let the world be conquered."

"And by whom?" asked George Whitley. "The Martians? Or the well-known little green men from Venus?"

"Neither of them. He was collaborating with someone nearer home. You'll realize who I mean when I tell you he was a myrmecologist."

"A which-what?" asked George.

"Let him get on with the story," said Drew, from the other side of the bar. "It's past ten, and, if I don't get you all out by closing time *this* week, I'll lose my license."

"Thank you," said Harry with dignity, handing over his glass for a refill. "This all happened some years ago, when I was on a mission in the Pacific. It was rather hush-hush, but, in view of what's happened since, there's no harm in talking about it.

"Three of us scientists were landed on a certain Pacific atoll, not a thousand miles from Bikini, and given a week to set up some detection equipment. It was intended, of course, to keep an eye on our good friends and allies, when they started playing with thermo-nuclear reactions—to pick some crumbs from the A.E.C.'s table, as it were. The Russians, naturally, were doing the same thing, and occasionally we ran into each other. Then both sides would pretend that there was nobody here but us chickens.

"This atoll was supposed to be uninhabited, but this was a considerable error. It actually had a population of several hundred million—"

"*What!*" gasped everybody.

"—several hundred millions," continued Purvis calmly, "of which number, one was human. I came across him when I went inland one day, to have a look at the scenery."

"Inland?" asked George Whitley. "I thought you said it was an atoll. How can a ring of coral—"

"It was a very plump atoll," said Harry firmly. "Anyway, who's telling this story?" He waited defiantly for a moment, until he had the right of way again.

"Here I was, walking up a charming little river-course underneath the coconut palms, when, to my great surprise, I came across a waterwheel—a very modern-looking waterwheel, driving a dynamo. If I'd been sensible, I suppose I'd have gone back and told my companions, but I couldn't resist the challenge, and decided to do some reconnoitering on my own. I remembered that there were still supposed to be Japanese troops around, who didn't know that the war was over, but that explanation seemed a bit unlikely.

"I followed the power-line up a hill, and there, on the other side, was a low, whitewashed building, set in a large clearing. All over this clearing were tall, irregular mounds of earth, linked together with a network of wires. It was one of the most baffling sights I have ever seen, and I stood there and stared for a good ten minutes, trying to decide what was going on. The longer I looked, the less sense it seemed to make.

"I was debating what to do, when a tall, white-haired man came out of the building and walked over to one of the mounds. He was carrying some kind of apparatus and had a pair of earphones slung around his neck, so I guessed that he was using a Geiger counter. It was just about then that I realized what those tall mounds were. They were termitaries—skyscrapers, in comparison to their makers, far taller than the Empire State Building, in which the so-called white ants live.

"I watched with great interest, but complete bafflement, while the elderly scientist inserted his apparatus into the base of the termitary, listened intently for a moment, and then walked back towards the building. By this time, I was so curious that I decided to make my presence known. Whatever research was going on here obviously had nothing to do with international politics, so I was the only one who'd have anything to hide. You'll appreciate later just what a miscalculation *that* was.

"I yelled for attention and walked down the hill, waving my arms. The stranger halted and watched me approaching, he didn't look particularly surprised. As I came closer, I saw that he had a straggling mustache that gave him a faintly Oriental appearance.

"He was about sixty years old, and carried himself very erect. Though he was wearing nothing but a pair of shorts, he looked so dignified that I felt rather ashamed of my noisy approach.

" 'Good morning,' I said apologetically. 'I didn't know there was anyone else on this island. I'm with an—er—scientific survey party over on the other side.'

"At this, the stranger's eyes lit up. 'Ah!' he said, in almost perfect English. 'A fellow scientist! I'm very pleased to meet you. Come into the house.'

"I followed gladly enough—I was pretty hot after my scramble —and I found that the building was simply one large lab. In a corner were a bed and a couple of chairs, together with a stove and one of those folding washbasins campers use. That seemed to sum up the living arrangements. But everything was very neat and tidy. My unknown friend might be a recluse, but he believed in keeping up appearances.

"I introduced myself first, and, as I'd hoped, he promptly responded. He was one Professor Takato, a biologist from a leading

Japanese university. He didn't look particularly Japanese, apart from the mustache I've mentioned. With his erect, dignified bearing, he reminded me rather of an old Kentucky colonel I once knew.

"After he'd given me some unfamiliar but refreshing wine, we sat and talked for a couple of hours. Like most scientists, he seemed happy to meet someone who could appreciate his work. It was true that my interests lay in physics and chemistry, rather than on the biological side, but I found Professor Takato's research quite fascinating.

"I don't suppose you know much about termites, so I'll remind you of the salient facts. Termites are among the most highly evolved of the social insects, and live in vast colonies throughout the tropics. They can't stand cold weather nor, oddly enough, can they endure direct sunlight.

"When they have to travel from one place to another, they construct little covered roadways. They seem to have some unknown and almost instantaneous means of communication, and, though the individual termites are pretty helpless and dumb, a whole colony behaves like an intelligent animal.

"Some writers have drawn comparisons between a termitary and a human body, which is also composed of individual living cells, making up an entity much higher than the basic units. The termites are often called 'white ants', but that's a completely incorrect name. They aren't ants at all, but quite a different species of insect—or should I say 'genus'? I'm pretty vague about this sort of thing.

"Excuse this little lecture, but, after I listened to Takato for a while, I began to get quite enthusiastic about termites myself. Did you know, for example, that they not only cultivate gardens, but also keep cows—insect cows, of course—and milk them? Yes, they're sophisticated little devils, even though they do it all by instinct.

"But Id better tell you something about the Professor. Although he was alone at the moment, and had lived on the island for several years, he had a number of assistants, who brought equipment from Japan and helped him in his work. His first great achievement was to do for the termites what von Frische had done with bees—he'd learned their language. It was much more complex than the system of communication that bees use, which, as you probably know, is based on dancing.

"I understood that the network of wires linking the termitaries to the lab not only enabled Professor Takato to listen to the termites talking among themselves, but also permitted him to speak to them. That's not really as fantastic as it sounds, if you use the word "speak" in its widest sense.

"We speak to a good many animals—not always with our voices, by any means. When you throw a stick for your dog and expect him to run and fetch it, that's a form of speech—sign language. The Professor, I gathered, had worked out some kind of code which the termites understood, though how efficient it was at communicating ideas I didn't know.

"I came back every day, when I could spare the time, and, by the end of the week, we were firm friends. It may surprise you that I was able to conceal these visits from my colleagues, but the island was quite large, and we each did a lot of individual exploring. I felt, somehow, that Professor Takato was my private property. I did not wish to expose him to the curiosity of my companions. They were rather uncouth characters—graduates of some provincial non-specialized universities like Oxford or Cambridge.

"I'm glad to say that I was able to give the Professor a certain amount of assistance, helping him fix his radio and line up some of his electronic gear. He used radioactive tracers a good deal, to follow individual termites around. He'd been tracking one with a Geiger counter when I first met him, in fact.

"Four or five days after we met, his counters began to go haywire, and the equipment we'd set up began to reel in its recordings, Takato guessed what had happened—he'd never asked me exactly what I was doing on the island, but I think he knew. When I greeted him, he switched on his counters and let me listen to the roar of radiation. There had been some radioactive fall-out—not enough to be dangerous, but enough to bring the background way up.

"'I think,' he said softly, 'that you physicists are playing with your toys again. And very big ones, this time.'

"'I'm afraid you're right,' I replied.' We couldn't be sure until the readings had been analyzed, but it looked as if Teller and his team had begun the hydrogen reaction. 'Before long, we'll be able to make the first A-bombs look like damp squibs.'

"'My family,' said Professor Takato, without any visible emotion, 'was at Nagasaki.'

"There wasn't a great deal I could say to that, and I was glad when he went on to add, 'Have you ever wondered who will take over when we are finished?'

"'Your termites?' I countered, half facetiously. He seemed to hesitate for a moment. Then he said quietly, 'Come with me—I have not shown you everything.'

"We walked over to a corner of the lab, where some equipment lay concealed beneath dust-sheets, and the Professor uncovered a rather curious piece of apparatus. At first sight, it looked like one of the manipulators used for the remote handling of dangerously radioactive materials. There were handgrips that conveyed movements through rods and levers, but everything seemed to focus on a small box, a few inches long on each side.

" 'What is it?' I asked.

" 'It's a micromanipulator. The French developed them for biological work. There aren't many around yet.'

"Then I remembered. These were devices with which, by the use of suitable reduction gearing, one could carry out the most incredibly delicate operations. You moved your finger an inch—and the tool you were controlling moved a thousandth of an inch.

"The French scientists who had developed this technique had built tiny forges, on which they could construct minute scalpels and tweezers from fused glass. Working entirely through microscopes, they had been able to dissect individual cells. Removing an appendix from a termite—in the highly doubtful event of the insect possessing such an organ—would be child's play with such an instrument.

" 'I am not very skilled at using the manipulator,' confessed Takato. 'One of my assistants does all the work with it. I have shown this to no one else, but you have been very helpful. Come with me, please.'

"We went out into the open, and walked past the avenues of tall, cement-hard mounds. They were not all of the same architectural design, for there are many different kinds of termites —some, indeed, don't build mounds at all. I felt rather like a giant walking through Manhattan, for these were skyscrapers, each with its own teeming population.

"There was a small piece of metal—not wooden, for the termites would soon have fixed that—beside one of the mounds. As we entered it, the glare of sunlight was banished. The Professor threw a switch, and a faint red glow enabled me to see various types of optical equipment.

" 'They hate light,' he said, 'so it's a great problem observing them. We solved it by using infrared. This is an image-converter of the type that was used in the war for operations at night. You know about them?'

" 'Of course,' I said. 'Snipers had them fixed on their rifles, so they could go sharpshooting in the dark. Very ingenious things—I'm glad you've found a civilized use for them.'

"It was a long time before Professor Takato found what he wanted. He seemed to be steering some kind of periscope, probing through the corridors of the termite city. Then he said, 'Quickly —before they've gone!'

"I moved over and took his position. It was a second or so before my eye focused properly, and longer still before I understood the scale of the picture I was viewing. Then I saw six termites, greatly enlarged, moving rather rapidly across the field of vision. They were traveling in a group, like the huskies forming a dog-team. This was a very good analogy, because they were towing a sledge.

"I was so astonished that I never even noticed what kind of load they were moving. When they had vanished from sight, I turned to Professor Takato. My eyes had now grown accustomed to the faint red glow, and I could see him quite well.

" 'So that's the sort of tool you've been building with your micromanipulator!' I said. 'It's amazing! I'd never have believed it.'

" 'But that is nothing,' replied the Professor. 'Performing fleas will pull a cart. I haven't told you what is so important. We only made a few of those sledges. *The one you just saw they constructed themselves.*'

"He let that sink in. It took some time. Then he continued quietly, but with a kind of controlled enthusiasm in his voice,

'Remember that termites, as individuals, have virtually no intelligence. But the colony as a whole is a very high type of organism— an immortal one, barring accidents.

" 'It froze in its present instinctive pattern millions of years before Man was born, and, by itself, it can never escape from its present sterile perfection. It has reached a dead-end—because it has no tools, no effective way of controlling nature.

" 'I have given it the lever to increase its power, and now the sledge to improve its efficiency. I have thought of the wheel, but feel it best to let that wait for a later stage—it would not be very useful now. The results have exceeded my expectations. I began, with the termitary alone—but now they all have the same tools. They have taught each other, and that proves they can co-operate. True, they have wars—but not when there is enough food for all, as there is here.

" 'But you cannot judge the termitary by human standards. What I hope to do is to jolt its rigid, frozen culture—to knock it out of the groove in which it has stuck for so many millions of years. I will give it more tools, more techniques—and, before I die, I hope to see this organism beginning to invent things for itself.'

" 'Why are you doing this?' I asked, for I knew there was more than mere scientific curiosity here.

" 'Because I do not believe that Man will survive, yet I hope to preserve some of the things he has discovered. If he is to be a dead-end, I think another race should be given a helping hand. Do you know why I chose this island? It was so that my experiment should remain isolated. My supertermite, if it ever evolves, will have to remain here until it has reached a very high level of attainment. Until it can cross the Pacific, in fact . . .

" 'There is another possibility. Man has no rival on this planet. I think it may do him good to have one. It may be his salvation.'

"I could think of nothing to say. This glimpse of the Professor's dream was so overwhelming—yet, in view of what I had just seen, so convincing. For I knew that Professor Takato was not mad. He was a visionary, and there was a sublime detachment about his outlook, but it was based on a secure foundation of scientific achievement.

"It was not that he was hostile to mankind—he was sorry for Man. He simply believed that humanity had shot its bolt, and wished to save something from the wreckage. I could not feel it in my heart to blame him.

"We must have been in that little hut for hours, exploring possible futures. I remember suggesting that, perhaps, there might be some kind of mutual understanding, since two cultures so utterly dissimilar as Man and Termite need have no cause for conflict. But I couldn't really believe this, and, if a contest comes, I'm not certain who will win. For what use would man's weapons be against an intelligent enemy who could lay waste all the wheatfields, all the rice-crops in the world?

"When we came out into the open once more, it was almost dusk. It was then that the Professor made his final revelation.

" 'In a few weeks,' he said, 'I am going to take the biggest step of all.'

" 'And what is that?' I asked.

" 'Cannot you guess? I am going to give them fire.'

"Those words did something to my spine. I felt a chill that had nothing to do with the oncoming night. The glorious sunset beyond the palms seemed symbolic—and, suddenly, I realized that the symbolism ran even deeper than I had thought.

"That sunset was one of the most beautiful I had ever seen, and it was partly of man's making. Up there in the stratosphere, the dust of an island that had died this day was encircling the earth. My race had taken a great step forward—but did it matter now?

" *'I am going to give them fire.'* Somehow, I never doubted that the Professor would succeed. When he had done so, the forces that my own race had just unleashed would not save it.

"The flying boat came to collect us the next day, and I did not see Takato again. He is still there, and I think he is the most important man in the world. While our politicians wrangle, he is making us obsolete.

"Do you think that someone ought to stop him? There may still be time. I've often thought about it, but I've never been able to think of a really convincing reason for my interfering. Once or twice, I nearly made up my mind, but then I'd pick up the newspaper and see the headlines.

"I think we should let them have the chance. I don't see how they could make a worse job of it than we've done."

MIKE SHAYNE'S LATEST BOOK

"Weep for a Blonde Corpse"

by Brett Halliday

●

Science fiction readers seldom can be kept away from the best in mystery fiction. They'll follow a long trail, if need be, to seek it out. But now the trail is short indeed, and the best is right at hand in the current issue of SATELLITE'S *monthly companion on the high road to peak reading pleasure excitement,* MICHAEL SHAYNE MYSTERY MAGAZINE. *And the biggest thrill of all is a brand new Mike Shayne novel, starting with this issue and continuing on through two more generous installments. If you don't want to miss a single breathless, dropped heartbeat of Shayne's fierce intanglement with blondes and bullets and a false accusation of murder the time to hop on the bandwagon is now, immediately.*

Something Odd, Something Dangerous?—by

JOHN VICTOR PETERSON

The general was a stranger at the H-bomb test—

but so was the object that hovered close overhead!

FOOD FOR THE VISITOR

IT WAS H-HOUR minus ten minutes at Yucca Flat, Nevada. I lay in a slit trench, five miles from ground zero, wishing I were back with Colonel Kitchell at Battalion Headquarters—a concrete blockhouse—or, better still, seventy-five miles away in a Las Vegas bar.

Things were too snafu for my taste. The weather had been bad for days. There was a high overcast now. To top things, with M-minute coming up, my walkie-talkie had quit. I was stymied: I'm a line sergeant, not an electronics tech. The colonel, waiting to give orders and receive my reports, wouldn't take kindly to my oversight in neglecting to bring a tech along.

I lay there, getting more nervous by the second, alternating my gaze between my wristwatch and the faraway steel tower holding the bomb, and wishing the unidentified flying object would come around to liven things up.

I peered up at the clouds and, completely unexpected, a heavy hand fell on my shoulder. I'm glad my nerves weren't too taut. At zero, I don't know what I would have done.

As it was, I cursed my platoon silently for not warning me and slowly twisted around nearly to rub noses with a prone and apparently triumphant individual whose helmet bore a brigadier general's star.

"Alertness is important, Sergeant," he said sharply. "We're simulating combat conditions, you know."

"I think I'm covered sir," I said, trying not to sound flippant as I hopefully looked past him. Fortunately, Corporals Herrmann and Zuewski had kept alert. Both had their submachineguns at the ready.

The general twisted around, spluttered, and said, "I saw your men, of course!"

I don't argue with officers.

"Your name, Sergeant?"

"Parker, sir—Wesley Parker."
"And your part in the mission?"
"Guinea pig," I told him, saying what I felt. "Five miles from ground zero. Medium H-bomb. We expect total destruction up to four miles. Not that there's much on this desert to destroy except humans. If the blast doesn't get us, the fallout will—unless the wind shifts, of course. Weather forecast a dead calm, but we've a twenty mile wind straight in the face. The walkie-talkie just fizzed, so we couldn't get withdrawal orders, even if there were any."

I hoped he'd suggest we withdraw, but he simply grunted, bit his lower lip and looked up at the overcast. Then a strange expression came upon his face.

"Damnably strange reflection, that," he said. "But with no break in the clouds, it couldn't be a reflection. Searchlight, I guess . . ."

I looked up, too, and chuckled.
"That's no searchlight, sir," I explained. "That's our UFO friend."

"UFO friend?"

I already suspected he was new to the area.

"We call it the UFO of Yucca," I said patiently. "It's been circling over the Flat, ever since they brought the bomb up from Los Alamos last week. I've been watching it every time it's been around and, funny thing, it's getting dimmer! Project reports describe them as glowing, as if incandescent. This one's a weak sister, I'd say."

"But, damn it, Parker!" he burst out, "hasn't someone investigated it? This warrants cancellation of the test!"

"Why? There's no positive proof a UFO's unfriendly, or not a natural phenomenon. Anyway, you know more about this than I!"

He winced. "I'm afraid I'm a Johnny-come-lately here, Parker. My last physical *was* my last, as far as the army's concerned. This is an old soldier's—my holiday. Although General Steuerwalt cleared me into the area, I'm afraid I wasn't fully briefed."

I'd been watching him as he talked and recognized him—James "Bull" O'Brien, a hero in the Phillipines and, later, in Korea. We had heard a lot about him, in our training at Fort Benning. I felt sorry for him now, gallant old infantryman that he was.

"Incandescent," he murmured.

I was momentarily dumbfounded. "Oh, the saucer, you mean, sir! Well, they've chased it with XF-150 Sunbeaters at two thousand m.p.h. and weren't even getting close until yesterday. Then they were gaining, until it took evasive action into clouds. Strange, but it keeps coming back, as if waiting for the big bang."

General O'Brien's gaze was quizzical. "Maybe it is. Observing—checking on our weapons progress."

He pursed his lips, thinking, silent. We watched the UFO flutter dimly against a dark cloud mass.

I looked at my watch. One minute to zero!

The general was wholly absorbed with the UFO.

I shook the walkie-talkie, and Colonel Kitchell's voice abruptly came from it. " . . . above all, use your protective glasses. Do not look directly at the blast. Keep completely below blast level. The countdown will begin in seconds."

General O'Brien abruptly decided that my slit trench was big enough for both of us. We weren't breathing very deeply anyway. The countdown had begun.

" . . . three, two, one, *zero!*"

The UFO pulsed dimly against the Nevadan sky.

The dark afternoon turned into a seething maelstrom of pink, yellow, crimson, orange and purple, the ground-blast a pink-yellow-pink sandwich, a crimson shaft suporting the crimson roiling fireball shot with orange and yellow, purple streamers darting spaceward . . .

The UFO was diving through the deep purple at the fireball, a fluttering moth courting a cosmic flame, then recoiling, agleam with new light, flinging itself spaceward with far more incredible speed than the climbing mushroom cloud.

"General," I cried, "it's brighter, *stronger!*"

"Yes," he said, "it *was* weak before, apparently too exhausted from its trip here to leave until it had *fed*. Now it has gone home."

"Home?" I asked, not puzzled but seeking his accord.

"Home to the sun," he said, "where else?"

THE SCIENCE FICTION COLLECTOR

Lively news about timely books and other pertinent science and fantasy information for the serious reader and avid collector.

by SAM MOSKOWITZ

TOMORROW AND TOMORROW by Hunt Collins. Pyramid Books, New York: 190 pages, $.35 . . . Hunt Collins is Evan Hunter. His definitive, best-selling novel of teen-age delinquency, *The Blackboard Jungle,* was made into an outstanding motion picture. Hunter is equally at home in the science fiction and mystery fields. A good many science fiction readers will recognize this novel as an elaboration of Hunter's memorable magazine story *Malice in Wonderland.* If we are to categorize the book, it follows in the tradition of Huxley's *Brave New World,* with the influence of Bester's *Demolished Man* strong upon it. The story follows the general pattern of the current popular dope addiction, alcoholism and huckstering novels.

Hunter's future world legalizes the taking of dope which is government-inspected, and it also sponsors a machine through which sensual delights can be seen, felt, heard and smelled. Some of its citizens even regard eating as a disgusting habit to be indulged in with the utmost privacy. All this takes place within an economic framework not unlike our contemporary one.

Hunter works hard for his realism. He even attempts to improvise slang of the future, which,

---

All books, special publications and science fiction news items indicated for review in this new column should be addressed to Sam Moskowitz, Science Fiction Collector, 127 Shephard Avenue, Newark 12, N. J.

though interesting, compels one to approach *Tomorrow and Tomorrow* with a little extra concentration. Typesetting tricks, odd snatches of verse, and technical rhetorical gimmicks are used rather unwisely at times for emphasis. He disregards the fact that his basic story elements are already so bizarre that the tricky prose effects should be used to soften—not to heighten—the strangeness.

Despite this, and overlooking the fact that today's trend has turned to science fiction of a somewhat different variety, *Tomorrow and Tomorrow* is a truly fascinating and eminently readable book.

NERVES by Lester Del Rey. Ballantine Books, New York: 153 pages, $.35 . . . This book cannot be properly reviewed without considering the author's history. Lester Del Rey rose to initial popularity by writing science fiction and fantasy with a predominant note of quite genuine sentimentality. His stories were characterized by a high degree of human emotion, which generated a real feeling of empathy between the reader and the lead characters.

There was the female robot who fell in love with a human; the intelligent dogs of the future who attempted to transform apes into the image of men whom they loved; the ancient god Pan who returned to find success in a modern jazz band; and the strange, bitter pride of the last Neanderthal man befriended by the Cromagnons. All are infused with the magic of credibility by Del Rey's special talent.

Nerves, on the other hand, is a complete departure from the type of story that brought Del Rey his early success. It is a tersely dramatic story, devoid of wishful thinking and stripped down to such a hard realistic core that at times it is impossible to tell where reality ends and fiction takes over. The scene is laid in an atomic energy conversion plant where an experimental product has gotten beyond control. The only man who can prevent destruction to the entire continent is buried in a lead coffin under a mass of radioactive debris. The fight to reach him, to save him from radioactivity, and eventually, to secure the continent against atomic disaster represents one of Del Rey's finest achievements. Readers familiar with the story in its original appearance will find the book doubled in length without any apparent loss of pace or story impact.

TALBOT MUNDY BIBLIO by Bradford M. Day. Science-Fiction and Fantasy Publications, 127-01 116th Avenue, South Ozone Park, 20, N. Y.: twenty-eight pages, mimeographed, $.50 . . . Talbot Mundy, a veteran, seasoned weaver of adventure tales for over thirty years, had a rare gift of the occult and an appeal similar to that of the late H. Rider Haggard. Few of his titles are in print and such famous tales as *Tros of Samothrace, Purple Pirate, Old Ugly Face,* and *Jimgrim* are avidly searched for by a coterie of Mundy devotees.

For such collectors, Bradford Day has rendered a most valuable service. The present bibliography, the most comprehensive to date of Mundy's works, lists his books and magazine stories and assembles an abundance of known data concerning them. In addition, associational articles accompany the material supplied by Dr. J. Lloyd Eaten and Mr. Day.

HIGHWAYS IN HIDING by George O. Smith. Gnome Press, New York: 223 pages, $3.00 . . . Through the years of his science fiction writing, George O. Smith has been pegged as a writer who stayed inspirationally close to the electronic laboratory for a good many of his ideas. He has even been accused of letting the science completely dominate the story. In *Highways in Hiding*—far and away the finest story he has ever written—Smith deserts the physical sciences for the world of extrasensory perception, and seems to have found his medium. Few science fiction novels in the past four or five years possess the pace and high level of interest maintained throughout this book. And I believe *Highways in Hiding* will definitely rank high in the International Fantasy Awards as one of the best fantasy books of the year.

THE CIRCUS OF DR. LAO *and Other Improbable Stories.* Edited by Ray Bradbury. Published by Bantam Books, New York: 210 pages, $.35 . . . If it is not already taken for granted, it should be reiterated that Ray Bradbury is as fine an anthologist as he is an outstanding author. It has taken Bradbury more than three years to find the time to compile his second anthology. His first, *TIMELESS STORIES FOR TODAY AND TOMORROW,* was published by Bantam Books in 1952, and proved Bradbury's ability to select fine, out-of-the-ordinary fantasies.

The Circus of Dr. Lao by Charles G. Finney, as a hard-cover book has had two American editions, and one in England. It is doubtful if more than five thousand copies were *printed* of the three combined offerings. Despite its limited sale, selective collectors of fantasy have long regarded the book as a modern masterpiece. Rereading this short novel, eleven years after the appearance of its last American edition in 1945, I find my present reaction entirely in accord with the one I expressed in a fan journal at the time:

"In the history of fantasy there has neven been a book quite like *The Circus of Dr. Lao*. Its extravagant fantasy, grounded deeply in the realism of Abalone, Arizona, where the action transpires, is utterly different. You will find here and there a trace of the whimsey of Thorne Smith, the modernized mythology of L. Sprague de Camp and Fletcher Pratt, the symbolism of Cabell, and the naughty provocativeness of any number of authors. But most of all there will be a great deal of Charles Finney, and Finney is Finney and not a style copiest.

"One cannot outline the plot, because there is no plot, nor is there a single philosophy, but rather a variety of philosophies which the author presents for comparison. To tell you that the tale is about a strange circus run by a Chinaman, starring some of the most fantastic creatures of mythology, is to make it sound trite, and *The Circus of Dr. Lao* may be called anything but trite."

In 1945 I said: "The price, five dollars, is high, but the selective collector, searching for only a few books of exceptional quality, should not hesitate to pay." What can I add now that the price is only $.35—and eleven short stories, from distinguished sources, have been generously tossed in as a bonus?

WATCH FOR OTHER *ARTHUR C. CLARKE* STORIES IN FUTURE ISSUES

THE ATTIC VOICE

Sam Dundy thought his father's life was an open book—but that was before he found the strange transmitter, after Dad died.

By ALGIS BUDRYS

THE HEAVY oak-veneer desk in the attic had been the family catch-all for years. The pigeonholes under the jammed rolltop were crammed with unimportant letters, receipted bills, circulars from the Department of Agriculture and all the other papers that might, someday, have become important but never had.

Now Sam Dundy's father was dead, collapsed over his last straight furrow, the tired heart finally pushed too far, and Sam was home from State College. There were arrangements to be made, the estate to be put in order, and his mother was unable to bring herself to it. Sunk within himself, his emotions numb, Sam had been doing the best he could.

He had gone through the main records, down in the dining room desk, but there still might be something up here in the attic that he ought to know about—some record of a loan, or an insurance policy, that would mean another debt left behind by his father, or —Sam knew how little hope there was in this—some money due the estate.

He pulled open the bottom drawer, and lifted off a bundle of brittle yellow newspapers. Under these was a worn book, with a cheap red binding that had become faded and splotched. The imitation gold stamping was flaked and half-gone. Sam picked up the book and squinted at the title page in the dim light coming through the attic window. The attic wasn't wired.

With dusk falling, he should have brought a flashlight. But he had been doing things like that for the past two days—moving in shock, forgetting things, making mistakes, falling into long periods of mental drifting while he stopped wherever he was and simply day-dreamed.

The book was something called *Every Farmer's Home Cyclopedia of Simple Veterinary Science.* He'd never heard of it, but he wasn't surprised. It was the kind of book salesmen were continually selling on farms, for one dollar down and thirty-five cents a week for twenty weeks. He couldn't imagine his father spending the money, but, as he leafed through the pretentiously printed, cheap paper pages, he saw the heavy underlinings and notations in the margins, done in his father's hand with a pencil stub.

Sam looked at it, weighing it in his hand, slowly shaking his head. It was like handling a piece of the past—like knowing suddenly, without any doubt, exactly what was going on inside a man's head.

Here it was—the long-forgotten first sign of his father's resolve to have him go to veterinary school, perhaps the source of it. Sam looked at the sections that had drawn most of his father's attention.

A chapter on Animal Surgery had its pages smudged from what must have been incessant reading. Sitting in the near-dark, Sam had no trouble picturing his father, in the worn armchair under the standing lamp in the parlor, with the book in his lap and his blunt forefinger moving slowly under each line, his face set in a frown of concentration.

Lying on the table next to him would be the grammar school dictionary, most of its pages loose in the binding. At intervals, his father would hold his place with one finger while he leafed through the dictionary, either muttering the definition over and over until he had memorized it or else scowling in momentary anger when the dictionary failed to include the word that had puzzled him.

The annotations almost covered the type on some of the pages. In several places, there was a curt *Won't work.* Once or twice, he saw *Try it,* following an underscored paragraph, and he wondered what his father had been doing.

At the end of the chapter, his father had blocked in a characteristic comment—*Can't pay the price for the tools, anyway. Don't know enough to make some. Haven't got the brains to handle them right, if I did. Have to give this up and try something else.*

In the same way his father had talked to the plow and the water pump, his father had talked to the book.

Sam smiled in the dark. Every bit of work his father did was accompanied by a steady drive of short, declarative sentences. "Gotta lift you *over* that rock . . ." "Push this part *in* here . . ."

"Gonna fence you *this* Fall . . ."

So here was the record of something else planned, possibly attempted and angrily abandoned, because there were too many things against its achievement.

At this thought, Sam's smile faded. His father's life had been overfull of such defeats. The old, hard-jawed man had fought his losing battle every step of the way, trying with strength and hard work and stubborness to do what money would have accomplished.

Sam remembered his father's harsh snort. "If farmers had money, youngster, most of 'em wouldn't be farmers."

So here was another time his father hadn't been able to carry something through. It must have been a long time ago, because he couldn't remember it at all.

He closed the book and put it down on top of the newspapers. It would look strange, sitting on his bookshelf at college alongside the legitimate texts. But it would sit there, if he ever went back.

Sam looked down into the drawer. There was still something else in there, something that looked like a flat box in this light. He took it out, surprised to find that it was hard—not as cold as metal, but not as slickly smooth as plastic. He held it up to the fading light, but there was nothing to see except a faint seam, that marked off a lid, and a catch.

Frowning, he pushed at the little metal square, seeking to unfasten it as a pressure latch first, then trying to slide it open. It didn't work in either way. Finally, he found the hairline that divided it in two, and pushed in opposite ways. The latch opened, but, when he tried to lift the lid, it stayed locked. The catch had slipped back as soon as he let it go.

Annoyed at the designer, who seemed to have thought that people had three hands, he fumbled the lid up with his little fingers, while holding the catch open, and saw that the box was a portable radio.

He frowned at the three knobs, the dial, and the two-inch grill that were set into the surface of the box under the lid. The dial was glowing luminously, and he wondered what kind of paint had managed to store up light for so long. But he wondered more about the radio's being there at all.

There was an old Sears-Roebuck set downstairs, sitting in the parlor, with its veneer splitting from old age. But it played well enough to get the weather bulletins and the Farm and Home Hour. Why had his father spent the money on a late-model transistor set and hidden it away up here, in the attic?

He couldn't figure out what make it was, either. The dial was hard to read—it looked like some trick system, all symbols instead of numbers, which didn't make any sense at all. It might be a cheap Japanese set—something like that—from the looks of the symbols, but the Japs used regular American numbers on the ones they sold over here.

He tried to turn one of the knobs, and found that to do so required more strength than he had expected. It didn't seem to do anything, either. The next one moved the pointer on the dial. So the third one had to be the switch. He twisted it and felt the box begin to vibrate softly in his hands.

But nothing happened after he had waited a long minute, and he wondered if the first knob was a volume control that he had turned all the way down. He twisted it back toward its first position.

It had turned completely dark in the attic. There was still some light outside, but, toward the east, the stars were out, shining pure white in the purple-black sky.

There was a click as he reset the first knob.

Faraway, and thin, he heard: *"N'fera, n'feri, n'fero . . . n'fera, n'feri, n'fero . . . . d'anclaf, n'fera."*

There was a pause. And then, still far, still thin, but nearer, he heard the reply, in another voice as dry and rustling as the first had been soft and lilting. *"N'fera,*

socsim. Socsim. D'anclaf, n'feri."

Something frightened him. He reached out and closed the radio's lid, and sat looking up at the distant stars, wondering what language he had heard . . .

He rapped softly on his mother's door, his face troubled. He didn't want to bother her, but he had to get the radio business settled. He frowned as he tried to think of how he was going to ask her. If she didn't know anything about it, he didn't want to upset her any more.

"Who is it?"

His mother's voice was nervous and apprehensive.

"It's Sammy, Mom," he said patiently, though there was no one else it could have been. "May I talk to you for a minute?"

"Oh—Sammy. I wasn't asleep yet—just dozing. Were you upstairs?"

"Up in the attic, Mom," he reminded her.

"Looking through the old desk? Did you find anything?" she asked.

"I don't know, for sure. May I come in and talk to you?" He knew that, if he didn't remind her, she'd forget what he wanted.

"Yes, of course, Sammy." He heard her bedclothes rustle, and opened the door. There was no light on, and he blinked as he made his way to her bed and sat down on the chair beside it. He took her hand.

"How are you, Mom?" he

asked gently. He could see the paleness of her face in the dim light, and as his eyes slowly became accustomed to it, he saw that she was keeping her eyes closed. Her hand was limp in his, and her breathing was shallow.

"Pretty well," she replied wanly. The fingers of her left hand twisted at the edge of the comforter. "I think I'll be able to get to sleep all right tonight. Is everything all right on the farm."

He nodded. "Everything's fine, Mom," he answered uncomfortably, remembering. "We've got to take good care of Annie," his father had often said, shaking his head angrily, with that frequent anger which came over him, and was coming over his son, now, at things he couldn't do anything about.

"I guess there's some women shouldn't ever have to farm," he had said slowly. "But what else can we do, boy? If a man's got a farm, he's got to work it. If a man's been a farmer all his life, he's got to stick with that. But you take care of her, if I can't. You're my hope, boy. You're the best hope I've got in this world."

"Did you find anything in the attic, Sammy?" his mother asked, forgetting again.

She hadn't been made for a life of never having enough, Sam thought to himself. She wasn't intended to live as she had had to for thirty years, out here in country, where winter froze you for half the year, and the summer baked the blood out of your body. Just one child to help her, and that one not a daughter but a son to work with his father—working at the very thing that ground her life down flat, too, and not able to work enough to make it pay, no matter how little you slept, or how your arms and back ached. It couldn't be done. Two men couldn't make it pay—they could only drag enough out of the land so they could hang on, year after year, slipping downhill by inches.

But you had the land, and you were born to it, and what else could you do? You hung on, hoping for you didn't know what. Going to school, grudging every hour of time, but knowing there wasn't any way out at all, without the schooling. Going on to the State College, because, if you didn't, then it would be one more generation of Dundys that bled their life into the ground, hoping for something better for the next.

Six years at Veterinary school, then, and two more to go, taking two semesters for every single semester's credits, because you had to work part-time for clothes and lodging and a little extra to send home.

Two years to go—Sam's lips tightened into a straight white line. There wasn't any money to pay a hired man to work the farm. Even if there had been, no hired man—no two men—would sweat and stretch themselves as his father had.

"I don't know, Mom," he replied finally. He still didn't know how to go about asking her. He didn't even know what the radio's being in the desk could mean. But his father must have had some good reason for having it.

"Do you know if anybody ever left anything with Dad for safekeeping?" he asked at last. "Tools, or equipment of some kind? Or maybe somebody paid off a loan by giving Dad something like a radio? Do you know if he ever did anything like that?"

"Did you find some papers?" His mother's hand tightened in his, and her voice grew with hope. "Does somebody owe Tom? Do you think it'll be enough so you can keep going to school?" Then her hand grew limp again.

"No," she whispered, "that couldn't be. There's no sense trying to wish something into being true."

She had been with Tom Dundy for thirty years, working beside him as much as she had it in her to, and more. She had drawn on faith in her man for strength, through all that time, and the habit was still there.

Sam didn't have any way of doing anything for her. But he found himself wishing she didn't still have enough hoping in her to be weaker every time it died.

"You don't remember Dad's having anything like that around the house, Mom?" he kept at her, sorry he had started and wanting to get it finished.

"No, Sammy, not ever."

"All right, Mom. I thought there might be something, but I guess I was wrong. I'll keep looking."

He slipped his hand free of hers, leaned forward and kissed her. "Good night, Mom. I'm pretty sure I can work something out."

"Good night, Sammy. Don't stay up too late."

He left her room quietly, closing the door softly behind him. You kept hoping, somehow—Tom Dundy had taught both of them that. It wasn't a blind thing. It was knowing, deep inside yourself, that a man's strength and a man's mind ought to be enough in this world, that, if you held out long enough, things had to turn your way. But it was hard, losing the man.

He and his mother had both lived with his father, following him, taking the way he picked. Now he was gone, and they were both lost for a while. But there'd be a way out. The world hadn't changed merely because one man was gone.

Sam came into his own room and looked at the radio on the dresser . . .

The crops grew, and the rain

came down. There was always food growing, always a roof to live under. They were strong factors—permanent things that were rooted in more than one man's life. It was just that these factors didn't seem useful any longer.

Why had his father had that radio? Sam went to it and opened it without knowing exactly why, except that he was trying to find something to do.

Far away—farther than the first voices he'd heard—there was a burst of crystalline chiming, and a fainter answer. He twisted the stiff dial, moving down the band, and whisps of voices grew to clarity, faded, and were gone again. *reinni ser—. . . . grut tagat vol "Ssthethannn . . . flahmit somahal . . . dundi, wat—"*

His fingers froze.

"Dundy! *Dundy!*" The voice was harsh and deep, like no voice he had ever heard speaking English before. "Dundy, answer me, please! What's happened to you?"

Somewhere, among those people who spoke along these bandwidths, someone else was lost and uncertain because Tom Dundy was dead . . .

It was late at night, but Sam didn't look at his watch to check the time. For several hours, now, he had been sitting on the edge of his bed, listening to the radio calling his father's name at intervals, in a voice that sounded like the note of some heavy machine crushing out a tunnel in the rock far underground.

Of all the people in the world, if you knew anybody, you knew your parents. You knew the sound and mood of every kind of footstep they could take, the meaning of every cough that cleared their throats before they spoke to you. You knew the touch of their hands and the rustle of their clothes.

He tried to picture his father, sitting and listening to that voice —answering it somehow—but he couldn't. Nothing he knew or remembered was any help to him.

His father was a man who had kept much to himself. If he had wanted to talk to somebody over that radio, without letting anybody know about it, there was always time during the day when he could have used it in privacy. But Sam couldn't think of any reason for his doing it.

He remembered something his father had told him once— "Youngster, it's no good talking about something you're doing. If it works out, then you got something better to talk about. If it don't, nobody knows you've been a damn fool but you."

That would explain his father's not telling him. But it didn't explain what the radio was all about.

It might have been going on for years. His father had seen too many seasons come and go and come back again, to act for tomorrow, or next month.

"Son, how about you going off to Vet school up at State College? Seems to me we could swing it."

He remembered that with absolute clarity. His father, twelve years ago, had turned to him at supper one night and said it, with the idea thought over in his head for nobody knew how long beforehand.

"State pays for your schooling—no trouble there. Be a little tight, with clothes and food to pay for, and us missing your help here, but it's worth it."

It was. It was worth the extra work his father had to go back to doing, an older man than the one he had been before Sam could do a man's work. A veterinary didn't have to kill himself over a plow for his food and his roof. A veterinary's mother could live in a house in town, with no more chores to do, and a veterinary's wife wouldn't dry up and grow into a thin, pale woman with creases in her forehead and the skin chapping on her hands until it cracked open.

So now Sam was three-quarters of the way to being a graduate veterinary. And Tom Dundy had killed himself to get him there. That had been twelve years in the planning.

Sam raised his head. The voice on the radio was saying something else.

"I know someone is listening. I repeat—I know someone is listening to me. It is imperative that anyone connected with Thomas Dundy of Minnesota reply to this message." The voice was still heavy, still deep, but now it was desperate.

Sam didn't move. He wasn't sure that he should—he knew he didn't want to.

"Please—anyone with information regarding the present whereabouts of Thomas Dundy—it is urgent that I hear from him immediately. If he is unavailable, please locate someone truly able to speak for him."

Sam got to his feet and stood over the radio, his fists clenched.

"To operate the transmitting apparatus, move the first button to the—the left. Repeat—to transmit, switch the first button to the left. Speak toward the grill. I will now wait two minutes for a reply."

Sam reached out toward the knob, wondering who it was that wasn't quite sure of left and right.

Then he snapped the knob over, because he didn't dare not do it.

"This is Sam Dundy," he said shakily. "Who are you?"

He waited, but there wasn't any answer until he realized he ought to switch the knob back.

"—dy's son? Are you Tom Dundy's son? The student?"

"Yes. Look, who are you?"

"I can't tell you that if your father is available. Is he all right?"

"He's dead, mister. I found the radio by accident. I don't know anything about this. If you want anything, you'd better explain."

"Dundy died? How long ago?"

"Three days."

There was an odd, unidentifiable noise before the voice answered. Then it said, "You are almost finished with your studies, aren't you?"

"It'll be two years yet, mister."

"Two *years!* I don't know whether I can . . . "A pause, then, "I'm sorry for you about your father. Will you help me?"

Sam clamped his jaws together. "Not unless I know who I'm helping and what this is all about."

"I'm sorry—I'm being hasty. But I have a very good reason. If I do not have help soon, I shall be dead, too. When I lost contact with your father, I . . . You understand? I lost my composure. I don't—that is, I'm not in a position to extend my trust to very many people.

"Your father found me a number of years ago, but he could not help me. I gave him the communicator so we could keep in contact, and he could advice me when circumstances had changed. I have slowed my metabolism to nearly the critical level, but if you cannot help me soon, it will be too late. I apologize for my . . . brusqueness, but I am also prey to emotions. I ask you again—will you help me?"

Sam looked down at the radio —the communicator. He knew his face was pale. He wiped his hand across his face, and his fingers were damp when they came away.

Outside, he knew, the stars were shining cold and clear on this moonless night, and he knew, too, that there were voices whispering across the night between the stars.

He took a deep breath. "I don't know," he answered slowly. "I'd have to see you. How far away are you?"

The voice made the odd sound again. It might have been a sigh. "Somewhere on your father's holding. I don't know where, in relation to your domicile. I came down at night, and too quickly . . . I will broadcast a beam of sound. Walk along it, and I will guide you. You are a student— you are aware that not all organisms must resemble one of your race—that intelligence and emotion may nevertheless be present?"

"I'll be coming."

"I thank you."

Quietly, Sam walked downstairs and got the flashlight. Holding the communicator, he stepped out into the yard, listening as the voice said, "I will start the beam now." A series of sharp clicks began to come from the speaker.

Walking slowly in the darkness, picking his way with the flashlight's help, Sam left the yard, sometimes having to go off the beam and find it again, when confronted by a building or a fence.

He figured out where he was headed almost immediately. There was a section of woods that he and his father had never cleared, or had ever intended to, what with so many other things to take care of.

He came down at night, Sam thought, and quickly. I wonder where he's from? I wonder which of the languages I heard is the one his people speak?

The earth under his feet felt different, somehow, than it had ever felt before—not as solid, yet better. He walked into the woods, threading his way between the trees.

"I hear your footsteps." The beam was cut off, and the voice spoke to him again. "Walk slowly. Look down."

Some of the trees were half-toppled, leaning against others, as though a windstorm had pushed them over. He must have walked through here a hundred times.

"Stop."

He looked down, and the dead leaves stirred, a few feet in front of him. He saw movement under them, something that looked like a sluggish rock, not too big. He saw a stir of thick arms that were like petrified tree roots, bending under enormous pressure.

"I have internal injuries," the voice said, booming and grating out from the ground, instead of from the communicator. Sam switched it off and shut the lid automatically. "I cannot dig out my ship and repair it, as I am. I have no access to my medical supplies or the interstellar transmitter. I have only my personal kits." The rock twisted itself sideward, and he saw an oval patch on its side, with a dark stain in its center.

"I was thrown clear. We are a tough people, and the ship is made of the same substance as the communicator case, but I cannot reach it. I'm at your mercy. If you help me, my bargain with your father will hold. I will tell you how to construct some of the circuits which make the communicator and the ship's propulsion fields possible. I understand such knowledge is negotiable here."

Sam nodded. "Yes, it is. There's a patent office, and companies that build electronic instruments. But I think I'd help you anyway." He thought about the old *Veterinary Cyclopedia*, with his father's stubborn notes. "I think I can do it. If not right now, *very* soon."

"Thank you. You are much like your father."

"I hope so, Mister," Sam answered, "I hope so!"

continued from Back Cover

$1.00 IS ALL YOU PAY FOR ANY 3 OF THESE GREAT BOOKS
when you join the Club
Each One Packed from Cover to Cover With Thrills of Top-Flight Science-Fiction

THE ASTOUNDING SCIENCE-FICTION ANTHOLOGY
A story about the first A-Bomb... written *before* it was invented! Plus a score of other best tales from a dozen years of *Astounding Science-Fiction* Magazine selected by editor John W. Campbell, Jr. (*Publ. ed. $3.95.*)

THE TREASURY OF SCIENCE-FICTION CLASSICS
World-renowned stories that have stood the test of time—by H. G. Wells, Jules Verne, Sir Arthur Conan Doyle, Aldous Huxley, Philip Wylie, Edgar Allan Poe, E. M. Forster, F. Scott Fitzgerald, etc. 704 pages. (*Publ. ed. $2.95.*)

NOW—THE BEST NEW SCIENCE-FICTION BOOKS FOR ONLY $1.00 EACH!

IMAGINE—ANY 3 of these full-size, brand-new science-fiction books—yours for just $1! Each is crammed with the science thrills of the future... written by a top-notch science-fiction author. An $8.65 to $12.40 value, complete in handsome, permanent bindings. Each month the SCIENCE-FICTION BOOK CLUB brings you the finest brand-new full-length books FOR ONLY $1 EACH (plus a few cents shipping charges)—even though they cost $2.50, $3.00 and up in publishers' editions! Each month's selection is described IN ADVANCE. You take ONLY those books you really want—as few as 4 a year.

SEND NO MONEY
Mail Coupon TODAY!

Take your choice of ANY 3 of the new books described here—at only $1 for ALL 3. Two are your gift books for joining; the other is your first selection. Mail coupon RIGHT NOW to:

SCIENCE-FICTION BOOK CLUB
Dept. SSF-2, Garden City, N.Y.

OMNIBUS OF SCIENCE-FICTION
43 top stories by outstanding authors ...tales of Wonders of Earth and Man ... of startling inventions ... visitors from Outer Space ... Far Traveling ... Adventures in Dimension ... Worlds of Tomorrow. 562 pages. (*Publ. ed. $3.50.*)

THE BEST FROM FANTASY AND SCIENCE-FICTION
(*Current Edition*) The cream of imaginative writing...selected from the pages of *Fantasy and Science-Fiction Magazine*. Tales of adventure in other worlds ... mystery, intrigue, suspense in future centuries! (*Publ. ed. $3.50.*)

THE REPORT ON UNIDENTIFIED FLYING OBJECTS
by Edward J. Ruppelt
At last! The first authoritative report on hitherto hushed-up facts about "flying saucers"... by a former Air Force expert who was in charge of their investigation. NOT fiction, but amazing FACT! (*Publ. ed. $4.95.*)

THE DEMOLISHED MAN
by Alfred Bester
Ben Reich had committed "the perfect murder"—except for one thing, the *deadly enemy* that followed Reich everywhere... A MAN WITH NO FACE! (*Publ. ed. $2.75.*)

THE EDGE OF RUNNING WATER
by William Sloane
Julian Blair had created a frightening yet amazing machine that would *prove immortality!* "Suspense... ingenuity, and excellent description." —N. Y. Times. (*Publ. ed. $3.00.*)

THE LONG TOMORROW
by Leigh Brackett
After the Destruction, the Bible-reading farmers ruled the country. But there was still one community of Sin in the land. Fascinating tale of two young boys' search for the Truth which lay in this town of Evil. (*Publ. ed. $2.95.*)

WHICH 3 DO YOU WANT FOR ONLY $1.00?

SCIENCE-FICTION BOOK CLUB
Dept. SSF-2, Garden City, New York

Please rush me the 3 books checked below, as my gift books and first selection. Bill me only $1 for all three (plus few cents shipping charges), and enroll me as a member of the Science-Fiction Book Club. Every month send me the Club's free bulletin, "Things to Come," so that I may decide whether or not I wish to receive the coming selection described therein. For each book I accept, I will pay only $1 plus shipping. I do not have to take a book every month (only four during each year I am a member)—and I may resign at any time after accepting four selections.

SPECIAL NO-RISK GUARANTEE: If not delighted, I may return all books in 7 days, pay nothing, and this membership will be cancelled!

☐ Astounding Science-Fiction Anthology
☐ Best from Fantasy & S-F
☐ Demolished Man
☐ Edge of Running Water
☐ The Long Tomorrow
☐ Omnibus of Science-Fiction
☐ Report on UFO's
☐ Treasury of Science-Fiction Classics

Name_____(Please Print)
Address_____
City_____Zone_____State_____
Selection price in Canada $1.10 plus shipping. Address Science-Fiction Club, 105 Bond St., Toronto 2. (Offer good only in U. S. and Canada.)

The Fiction House Press Replica Line is available at www.FictionHousePress.com

Milton Keynes UK
Ingram Content Group UK Ltd.
UKHW040704050124
435493UK00001B/222